U0483700

符号中国 SIGNS OF CHINA

中国结

CHINESE KNOT

"符号中国"编写组 ◎ 编著

中央民族大学出版社
China Minzu University Press

图书在版编目(CIP)数据

中国结：汉文、英文 /"符号中国"编写组编著. —北京：中央民族大学出版社，2024.9
（符号中国）
ISBN 978-7-5660-2336-0

Ⅰ.①中… Ⅱ.①符… Ⅲ.①绳结－手工艺品－介绍－中国－汉、英 Ⅳ.①TS935.5

中国国家版本馆CIP数据核字（2024）第016825号

符号中国：中国结 CHINESE KNOT

编　　著	"符号中国"编写组
策划编辑	沙　平
责任编辑	罗丹阳
英文编辑	邱　械
美术编辑	曹　娜　郑亚超　洪　涛
出版发行	中央民族大学出版社
	北京市海淀区中关村南大街27号　邮编：100081
	电话：（010）68472815（发行部）　传真：（010）68933757（发行部）
	（010）68932218（总编室）　　　　（010）68932447（办公室）
经销者	全国各地新华书店
印刷厂	北京兴星伟业印刷有限公司
开　　本	787 mm×1092 mm　1/16　印张：8
字　　数	111千字
版　　次	2024年9月第1版　2024年9月第1次印刷
书　　号	ISBN 978-7-5660-2336-0
定　　价	58.00元

版权所有　侵权必究

"符号中国"丛书编委会

唐兰东　巴哈提　杨国华　孟靖朝　赵秀琴

本册编写者

徐　雯

前言 Preface

中国结是中国特有的民间手工编结艺术，它以其独特的东方神韵和丰富多彩的形态展现了中国人民的智慧和深厚的文化素养。

中国人对线有着深刻的认识与理解，对以绳线编制的结也情有独钟。这小小的结，表现了中国人高超的技巧和惊人的创造力。由于岁月的侵

As the unique handmade Chinese folk art, Chinese knot is a knitting handicraft with diverse shapes mirroring the charm of oriental art, wisdom of Chinese people and their artistic attainment.

Chinese people have showed special preference to knitting strings so that they have made diverse Chinese knots by their amazing creativity and clever hands. Knots were widely used in ancient times when they acted as notes, accessories and emotional expression. As time goes by, those beautiful knots made in old times are gone but the design can be seen from carved stones, paintings and accessories. The depiction of diverse shapes and beautiful patterns has been passed down. What's more, the profound connotation of Chinese knots has been rooted in later generations.

A small handmade Chinese knot is

蚀，那些年代久远的美丽的结早已不复存在，但刻画在砖石上、描绘在画卷里、再现于装饰中、记载于文字间的各类结仍以其丰富的样式、巧妙的结构、优美的形态和深刻的内涵感染着我们。无论是结绳记事还是束衣佩玉，无论是点缀装饰还是寓意传情，都是古时用结之频繁及广泛的表现和见证。

　　一个小小的结，不仅仅是美和巧的展示，更是自然灵性和人文精神的表露，蕴藏着质朴、自然的情愫，蕴藏着人的亲和之质以及机器永远无法替代的人工之巧。

endowed with both beauty and spirituality. Chinese people prefer knots in rustic and natural style because any kind of mechanical production is no substitute for ingenious handmade ones.

目 录 | Contents

中国结文化
Culture of Chinese Knot 001

中国结的来历与发展
Origin and Development of Chinese Knot 002

各种各样的结
A Variety of Chinese Knots 023

结的寓意
Implied Meaning of Chinese Knot 037

中国结的应用
Usage of Chinese Knot 043

服饰之结
Knot on Costumes 044

装饰之结
Decorative Knot 056

礼仪节庆之结
Chinese Knot for Festivities 065

附录：几种中国结的编制技法
Appendix: Knitting Techniques of
Several Types of Chinese Knots 079

基本结
Elementary Knot 080

组合结
Group Knot 104

中国结文化
Culture of Chinese Knot

中国结在古时是人们传递文化、交流情感的重要载体。追求完整、圆满、和谐是中华民族的重要价值取向，也是中国传统民间艺术的审美追求。在这种审美思想的影响下，中国结产生、传承并被不断地创新光大，其寓意无不与生命、喜庆、圆满、幸福、长寿等有关，其造型、色彩、命名，以及制作、运用等，皆反映了人们对福、寿、喜、财、平安、健康等的追求。

Affected by Chinese value orientation and the aesthetic of traditional folk art, Chinese knot which symbolizes pursuit of unity, perfection and harmony acted as an important mean of cultural exchange and communication among ancient people. Under the influence of this aesthetic ideology, Chinese knot has been gradually innovated in design, colors, names, knitting and usage; however, the implied meaning of Chinese knot related to happiness, longevity and harmony as well as a reflection of people's pursuit for good health, success and wealth remains unchanged.

> 中国结的来历与发展

结是极平常而又极有意味的事物。说其平常，是因为有绳便有结，每个人都会打结，而且常要打结，它平常得简直不被人们注意。说其有意味，是因为结除了有束物、连接、固定之用外，还有传情达意和装饰美化的功能。在悠悠岁月中，结与人们的精神生活结下了不解之缘，乃至成为一种能够寄托丰富意义或无尽情思的装饰品、陈设品。

结之产生，由来已久。但究竟源于何时，尚难说清。不妨作这样一种推测：在远古时期，那些线形材料，如草、藤、麻、棕、竹、葛、枝条等，为原始人随手可得。当他们采来并拧扭、交叉，用于穿系、捆扎果实及猎物时，最原始的

> Origin and Development of Chinese Knot

Making a knot is a common practice but meaningful. Tying a knot seems to be a piece of cake; that is to say, where there is a rope, there is a knot. When it comes to meaningfulness, knots can be used not only for fastening, connecting and tying objects but also for decoration and conveying special emotions. Knots have close relations with people's spiritual life through all the years, so they have become meaningful decorations leaving endless affection with people.

Tying knots can be dated back long time ago but it was unknown when it started. Let's hereby start with a hypothesis. From time immemorial, primitive people twisted natural material such as grass, rattan, hemp, palm fiber, bamboo and branches to bind around their clothing and strap the fruit and

• 大溪文化彩陶上的绞绳纹摹绘图（新石器时代）
Rubbings of Twisted Rope Patterns on Painted Pottery of Daxi Culture (Neolithic Age)

编结就产生了。不妨再作一种推测：在北京周口店山顶洞人的遗址中，有骨针和各种带孔的骨、牙类饰物，这表明旧石器时代已出现用于缝纫和穿连的人为加工的绳线。当人们用这绳线穿带孔饰物及缝制衣物时，必然会产生原始的、简单的绳结。如果以上推测成立的话，那么可以说，结的历史几乎与人类文明的历史一样悠久。

结之用途早有文字记载，如《易·系辞下》："上古结绳而治，后世圣人易之以书契。"其意是说，人们最早是用绳索打结的方法来记事的，后来才用文字取而代之。据说，在结绳记事时代"大事大结其绳，小事小结其绳"，人们

prey. They could be the prototypes for the original ropes. Let's put forward another hypothesis. The primitive people lived at the cave of the Loong Bone Hill, Zhoukoudian, Beijing. Bone needles and such ornaments as animal teeth and various bones with drilled holes were discovered among the ruins. It suggested that the manmade thread and rope for sewing appeared during the Paleolithic Period. If above two hypotheses are admissible, it is safe to say that the history of using knots is as old as history of human civilization.

The history of using knots was recorded in *Interpretation* in *The Book of Change*. As noted in this book, human beings originally kept a record of events by knots in ancient times before written

用不同的结来辅助记忆各种事情。至于当时那些大结小结究竟是什么样的，很难得知了，但从中我们可以了解到上古时期的绳结除了最简单的系扎、连接功用外，还有辅助记忆的功能。因此，绳结曾经是非常神圣的，它在当时人们的生活中起过重要作用。

在长期的社会实践中，由于绳结朴素的实用功能和神圣的记事功能，逐渐引起人们的审美关注，以至绳结日益增进着它的审美内涵。尽管缺乏实物，器物上的绳纹装饰也足以佐证绳结伴随人们的美的历

language was invented. It is said that an important event was marked by a large knot and vice versa. Various kinds of knots were used to help their memory of different events; however, it is hard to trace what they looked like. Anyway, we can learn that in addition to the basic function of binding and connecting objects, the knot also acted as a memory aid of recording the historical event in ancient times. In a word, the knot was regarded as a sacred object in ancient people's mind because it played an important role at that time.

After knots were widely used in daily life, people started to keep an eye on the design of knots so as to meet their higher aesthetic standards. Even though not a single physical knot was left from that time, from the ornamental patterns on the vessels we can know for sure how the knot design make people's life gracious and beautiful. During the Neolithic Age,

- 绳纹陶甗（yǎn）（夏）

甗是古代的一种蒸食用具，分为两部分，下半部是鬲（lì），用于煮水，上半部是甑（zèng），两者之间有镂空的箅（bì），用来放置食物，可通蒸汽。

Pottery Cookware with Ornamental Knot Pattern (Xia Dynasty)

This cookware was used for steaming food. It can be divided into two parts; water was poured into lower part and food was placed on the grate to be steamed.

• 错红铜绳结纹壶摹绘图（春秋）
Rubbing of Knots Pattern on Copper Pot (Spring and Autumn Period)

程。新石器时代陶器上丰富的绳纹装饰，表现出绳的各种纠缠、穿插、扭转之态。它们虽不是结形的模仿，却是结的精神和意象的写照。这种精神或意象最终向着审美的结饰方面发展，成为"绳结纹"。

there appeared diverse knot patterns such as twisting, winding and interweaving on pottery. Even though they were not exactly the same as the physical knots, they mirrored the abstract images. While an upsurge in aesthetic was in the making, the abstract images evolved into knot patterns accordingly and they were used for ornamental purpose.

结绳记事

在文字产生之前，远古人为了帮助记忆，采取过各种各样的记事方法，其中使用较多的是结绳和契刻。结绳记事即根据事件的性质、规模或所涉数量的不同结系出不同的绳结。简单的记数可用一根绳子打结，复杂的事件则可用多根绳子横竖交叉来表意。直到宋代，北方的游牧民族仍采用此法。

Using Knot to Keep a Record for Events

One of the most common ways to keep a record for events in ancient times was tying knots before the written language was invented. Considering different kind of the event, importance of the event and involved quantity, different kinds of knots would be used accordingly. Simple numbers were counted by means of knots tied on a single rope while complicated things had to be marked by a number of crossed ropes. The northern nomads still used this method until the Song Dynasty.

● 结绳记事
Using Knot to Keep a Record for Events

春秋战国时期，绳结不仅具有实用价值，近以相对独立的审美姿态进入装饰领域。例如，青铜双龙络纹瓿（bù）上作为装饰的络纹（绳结纹），虽然它的布局似乎都还保持着原来用绳子捆绑瓿、罐以便于提携的状貌，但毕竟是以装饰的形式展现在青铜器上。在这里络纹与双龙纹结合成富有特定文化意味和时代审美情趣的图案。这表明绳结所具有的实用功能和工匠们对绳结的深厚情感已融合发展成独到的装饰构思和贴切自然的器物装饰。在以后的装饰中，这种有趣的形式甚至被固定为一种装饰格式。

结不仅用于器物的装饰，同时

The knot became less functional but tended to be more ornamental during the Spring and Autumn and Warring States periods. Taking the bronze wine vessel with ornamental knot pattern as an example, it had a handle and remained the main structure of small jar but its design has combined the knot pattern with the double-loong pattern, which presented aesthetic taste and cultural significance of that era. Because the craftsman had preference for the knot, its decorative value became all the more prominent. Since then, the knot pattern has become a common ornament on utensils.

The knot also acted as accessories in ancient times. "Loose costumes with straps" was the fashion of that era. To

• 青铜双龙络纹瓿【局部】四角上的绳结纹摹绘图（春秋）
Rubbing of Duoble-loong and Rope Knots Patterns on Bronze Urn [Part] (Spring and Autumn Period)

也应用在人们的衣服和佩饰上。古人着装习尚"宽衣博带",要想使衣服贴体、保暖,就得靠带子来扎系并要打结。因此在古人衣装上,结的样式很丰富,有束服之结,也有装饰之结。飘逸的带与美妙的结已经成为美用一体的中国古典服装的重要组成部分。由于打结、解结是经常的事,所以古人身上常常佩有觿(xī)这种专门用于解结的工具。《说文》中解释道:"觿,佩角,锐端,可以解结也。"

穿衣要打结,戴帽、挂印要打结,佩玉也要打结。中国人佩玉风尚源远流长,这不仅仅因为玉质地温润,更因为它被赋予了种种文化

keep warm and clothes fitting closely but comfortably, ancient people knotted the strap; thereby the style of knot became very rich. Used for tying and decoration, the knot has constituted an important part of Chinese classical clothing. As the knot became a common practice, ancient people often had a tool for unknotting with them. The book *Explaining Characters* states: "*Xi*, pendant horn, with tapering tip, is used for untangling the knots."

The knot was closely connected with ancient costumes because it was often tied on dress, on hats, on seals and on jade pendants. For Chinese people, wearing jade accessories has long history. When it comes to jade, Chinese people

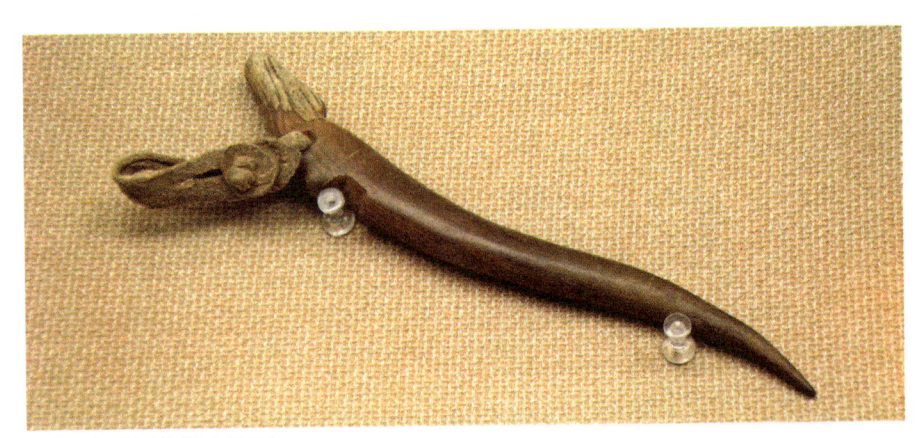

• 古代解结用的觿(春秋)
An Ancient Tool for Unknotting (Spring and Autumn Period)

内涵，具有象征性的精神品格。佩玉标志着一个人的身份地位、道德修养、品格情操，具有"载道""比德""达礼""显贵"的人文价值，故而历来有"君子必佩玉"（出自《礼记》）的讲究。

佩玉要借助绳带（古时玉饰上有孔），用绳带必然要打结。从已掌握的形象资料来看，早在战国时期，就已经存在较多样的佩玉绳结了。如河南信阳楚墓出土的彩绘木俑，上面清晰地描绘了当时人们佩玉打结的情形。另外，那些单耳结和下端的带饰，除了起系扎、固定玉璜及玉环的作用外，更有陪衬和装饰的功用。

often think of its symbolic spiritual connotations. Because beautiful jade has been endowed with various meanings related to Chinese culture, wearing jade accessories was often associated with a person's social status, power, personal ethic and morals. As noted in *The Book of Rites*, every gentleman in ancient times had a jade accessory with him.

There was a hole on a jade pendant to tie with silk braid. Referring to some related visual data, we can see a variety of silk braid widely used as early as the Warring States Period. Unearthed painted wooden figurines from the tomb of the State of Chu, in Xinyang, Henan Province vividly depicted ancient people wore jade pendants with silk braid during

- 河南信阳楚墓出土的组佩彩绘俑复原图

  图中二人身穿交领右衽直裾袍，宽袖，腰悬穿珠、玉璜、玉璧、彩结、彩环组佩。

  Restored Painting of Painted Wooden Figurines from the Tomb of the State of Chu, in Xinyang, Henan Province

  Two figures in the drawing wore crossed collar and right-buttoned garment with loose arced sleeves. It was decorated with diamond-shaped edge, beads, colored ribbons and groups of jade ornaments.

• 铜鼎上颇似绳结的蟠蛇纹（春秋）
Bronze Cauldron with Curled-up Serpent Pattern (Spring and Autumn Period)

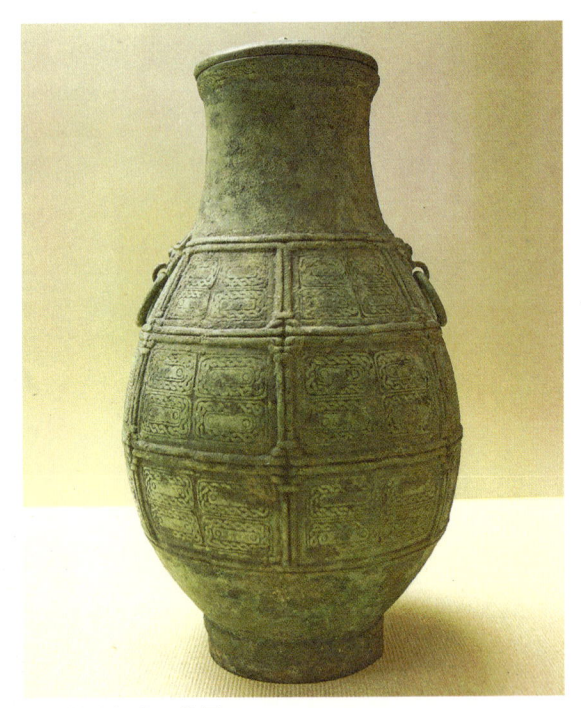

• 绳结纹铜壶（战国）
Bronze Pot with Rigging-shaped Pattern (Warring States Period)

结具有诸多的实用和装饰功能，同时它能被人们赋予一定的象征意义，因而具有招福纳祥、传情达意的作用。例如，汉代（公元前206—220）瓦当石刻中将龙蛇盘成结状，就具有浓厚的神秘色彩，这很可能是当时谶纬之学盛行的反映。而后来铜镜、砖刻、石雕上的结饰则日趋明快活泼，充满吉祥的寓意和欢乐的情绪。

the Spring and Autumn and the Warring States periods. Additionally, knots and silk braid also served as an ornament and a foil to semi-annular jade pendants and jade rings.

In addition to usefulness and decoration, the knot was often given a certain symbolic meaning so it was used to convey people's good wishes, such as bringing happiness and good fortune. Because of the prevalence of divination combined with mystical Confucianist belief, the craftsman of eaves tile carving during the Han Dynasty (206B.C.-220A.D.) took advantage of the image of curled-up loong and serpent to design the knot-shaped pattern full of mysterious atmosphere. Since then, more carved knot-shaped patterns appeared on bronze mirrors, bricks and stones but they tended to be lively, auspicious and joyful.

- 瓦当上的结纹（汉）
  Knot Pattern on Eaves Tile (Han Dynasty)

- 长沙马王堆漆勺柄上的绳结图案（汉）
  Lacquer Spoon with Knot Design on the Handle Unearthed from the Tomb of the Han Dynasty, Mawangdui, Changsha (Han Dynasty)

• 杂裾垂髾（shāo）服（魏晋南北朝）

杂裾垂髾服是魏晋南北朝时期女子的常见服饰，多在腰上系围裳，以结固定。

Zaju Chuishao Dress (Wei, Jin, Southern and Northern Dynasties)

It was a popular female dress of Wei, Jin, Southern and Northern dynasties, which was knotted around the waist by a long ribbon.

• 网结状树纹画像石（汉）
Stone Painting Depicting a Bark Texture Net (Han Dynasty)

唐宋是中国文化艺术发展的重要时期，结作为一种艺术装饰被大量运用于服饰和器物中。在流传至今的唐代铜镜中，就有一面双鸾衔绶纹铜镜，鸾鸟嘴衔飘动的系有"同心结"的绶带，寓意永结同心。唐代永泰公主墓的壁画中有一位仕女在腰带上佩结，形态为蝴蝶结。此外，唐代的《宫乐图》、唐代周昉的《挥扇仕女图》中也都有装饰结的身影。河南偃师酒流沟宋墓出土砖刻上的宋代女子，也在喇叭裤的外侧缀了一连串的结作为装饰。还有宋代的《货郎图》中也有装饰结

The Tang and Song dynasties witnessed an important development in art during which a large number of decorative knots were used on costumes and utensils. From the bronze mirror passed down from the Tang Dynasty, we can see the pattern of truelove knots. As shown in the pattern, mythical birds held ribbons tied with truelove knots in their beaks. In the Tomb of Princess Yongtai (Tang Dynasty), there was a wall painting, a court maid with a bowknot tied on her waistband. In addition, some decorative knot-shaped patterns were found on the paintings of the Tang

- 双鸾衔绶纹铜镜（唐）

Bronze Mirror Decorated with the Pattern of Mythical Birds Holding Ribbons Tied with Truelove Knots in Their Beaks (Tang Dynasty)

- 舞马衔杯镏金银壶（唐）

壶上的马颈部系有结饰，奔跑的姿态十分飘逸。

Gilt Silver Kettle with the Pattern of Running Horses (Tang Dynasty)

A knot was tied on the horse neck swaying with the pace of running.

● 乾陵永泰公主墓壁画（唐）

壁画中为首的是一位唐代贵族女子，梳高髻，身穿宽博的长裙，身旁有数位捧持生活用具的侍女，其中一侍女腰带上佩结。

Murals in the Tomb of Princess Yongtai (Tang Dynasty)

One noble woman was dressed in long and loose skirt wearing her hair high. Several maidservants held utensils, one of whom had a knot on her waist.

的形象。这一时期，结的实用性依然存在，而装饰性渐显突出，吉祥结、连环结等作为装饰形象广泛出现在服饰、器物、家具上，形态更加丰富，也更具内涵和寓意。

Dynasty, such as unearthed gilt silver censer, *A Court Maid Holding a Fan* painted by Zhou Fang and *Painting of Happy Party in the Imperial Palace*. Unearthed carved bricks from the Song-

宋代人物画中的结饰
Knot Patterns in the Figure Painting in the Song Dynasty

- 《大傩图轴》【局部】佚名（南宋）
Painting Scroll of Grand Ritual Ceremony [Part], by Anonymity (Southern Song Dynasty)

- 《维摩演教图卷》【局部】李公麟（北宋）
Painting Scroll of Vimalakīrti's Sutra [Part], by Li Gonglin (Northern Song Dynasty)

- 《闵予小子之什图卷》【局部】马和之（南宋）
Painting Scroll of the Tenth Eulogy by King Cheng of Zhou [Part], by Ma Hezhi (Southern Song Dynasty)

明清时期，中国的绳结技艺已发展至很高的水平。在诸多日常生活用品上，如轿子、窗帘、彩灯、帐钩、折扇、腰带、发簪、花篮、项坠、香囊、荷包及后来的眼镜袋、烟袋等等，都能见到美丽的花结装饰。此时还出现了灵活多变、一形多意的组合结，其样式繁多、

dynasty tomb in Yanshi, Henan Province showed decorative knot-shaped patterns on a lady's loose pants. *Painting of Peddlers* in the Song Dynasty depicted peddlers' clothing decorated with knots. During the Tang and Song dynasties, the knot as a decoration prevailed while its practical function still remained. Various auspicious knots were widely used on

• 砖雕人物服装上的结饰（元）
Knots on Brick Carving (Yuan Dynasty)

• 北京故宫中石狮子背后的结饰
Knot Pattern on the Back of Stone Lion in Forbidden City, Beijing

clothing, utensils and furniture, which enriched the design, connotation and implication.

The knitting skill of Chinese knot has reached a high level by the Ming and Qing dynasties. Beautiful knots were used to decorate a number of daily necessities, such as sedan chairs, curtains, lanterns, net hooks, fans, belts, hair clasps, flower baskets, pendants, perfume

• 法海寺妙音天女服饰上的结（明）
Ornamental Knots on Sarasvati's Costume in Fahai Temple (Ming Dynasty)

• 湘西民间家具雕刻中的绳结纹（近代）
Wood Carvings of Rope Knot Pattern on Furniture in the Western Area of Hunan Province (Modern Times)

配色考究、名称巧妙，实在令人目不暇接。

在漫长的岁月中，打结、编络子（清代称结饰品为"络子"）一直作为一项重要的女红手艺在民间广泛流传。往日，妇女们都会绾个扣、做个结、扎个穗子，就连老奶奶送给孙子的玩意儿上，也常常不忘编个结，以寄托祖辈对儿孙的祝福。至于一些礼品、定情物上的结就更是内涵丰富、意味深长了，诸如"同心结""盘长结""团锦结""蝴蝶结""双胜结"等等，不胜枚举。一根细绳，或一根自己加工的带子，在心灵手巧的妇女手中盘来绕去，不一会儿就成了一个漂亮的结子，而且往往同一种结有几种不同的编法，不同的结又经常是根据同一个思路做出来的，真可谓变化万千、妙趣无穷。

sachets and pouches. After glasses bags and tobacco pouches appeared, knots were attached for decoration. What's more, with the appearance of the group knot, decorative knots became more colorful, diverse and ingenious. Decorative knots were so exquisite that they took the fancy of people.

As one of the most important needlework skills, knotting and knitting have been passed down among people through the ages. It was no wonder that every woman was skillful in making knots and tassels. To send the best wishes, the elderly had their handmade knots attached to the present for their grandchildren. Ornamental knots on the present as a token of love were mostly meaningful. Such meaningful knots as truelove knot, *Panchang* knot, bouquet knot and bowknot were endowed with special affection. Following the design in mind, women who were handy with needlework made various knots right away; however, depending on their knitting technique of coiling, winding and twisting, all strikingly beautiful variants can be done by a thread or a strap.

• 门帘上的盘长结图案（近代）
Panchang Knot Pattern on Door Curtain (Modern Times)

结艺与女红

　　在中国古代，女子们在花结之中倾注了追求幸福的愿望，因此为它们冠以各种吉祥的名字。

　　清代出色的女红技艺是打络子，即编制花结饰品。《宫女谈往录》曾有记载："宫人出宫，都能带出一双巧手去……尤其出色的是打络子，满把攥着五颜六色的珠线、鼠线、金线，全凭十个手指头，往来不停地编织，挑、钩、拢、合，编成各种形象的图案，真是绝活。有时为了讨老太后的喜欢，把各种彩线拿来，用长针把线的一头钉在坐垫上，另一端用牙把主轴线咬紧、绷直，十个手指往来如飞，一会就编成一只大蝙蝠，和储秀宫门外往长春宫去的甬路上的活蝙蝠一模一样。"文中所写足以说明宫女们高超的结艺本领，而她们用金线来编结，可见人们对花结的重视。清代的金线制作有多种方法，一般是将金子做成的薄片切成细线，或织入线中，或绕包在线外。

- 宫廷如意上的结饰（清）
Decorative Knot on *Ruyi*, an S-shaped Ornamental Object Symbolizing Good Luck (Qing Dynasty)

Women's Needlework in Ancient Times

The desire for pursuit of happiness was seen from the auspicious names of various knots.

The skillful and outstanding needlework mastered by the court maid in the Qing Dynasty was making decorative knots. As recorded in *Memoirs of a Maid*, every court maid had a clever hand. Depending on their knitting skill, ten fingers, golden threads and colored beads, they were adept at making various knots. To please the empress dowager, Cixi, some maids showed this skill before her. They sat on cushions with one end of colored strings set on and the other end clenched in their teeth. Strings acted as a main axis. Their fingers were moving skillfully and a pattern of a large bat was done right away. It was as vivid as a real bat flying outside of the Palace of Gathered Elegance on the way to the Palace of Eternal Spring. According to the book, the maid used gold strings to make knots. It was thus clear that people took decorative knots seriously at that time. There were several ways to make gold string. After a sheet of gold was cut into filaments, they were either woven into string or wrapped the string.

- "万象更新"挂饰（清）

此挂饰以各种结串成，十分精致。

Decoration, the Design Implying that "Everything Takes on a Completely New Look" (Qing Dynasty)

Several kinds of knots were strung up on the exquisite perfume sachet.

今日，人们仍被精巧生动的中国结所吸引。中国联通（中国联合网络通信集团有限公司）以盘长结作为公司标志，象征着中国联通作为现代电信企业的井然有序、迅达畅通以及联通事业日久天长。2000年，中国把申奥标志定为与"结"神似的图案，使中国结愈加红火，受到越来越多人的关注和喜爱。如今，各式各样色彩鲜艳的中国结挂满了大街小巷，作为服饰配件和礼品包装、室内陈列装饰等，在现代人的生活中几乎无处不在。

• 靠垫上的结饰（现代）
Decorative Knots on the Back Cushion (Modern Times)

• 礼品盒上的结饰（现代）
Decorative Knots on the Gift Box (Modern Times)

Nowadays, people are still focused on the delicate and vivid Chinese knot. China Unicom (China United Network Communications Group Co., Ltd.) as a modern telecommunication enterprise uses *Panchang* knot as its logo symbolizing unimpeded communication and flourished business. In 2000, the logo of bidding for the host country of the Olympics was the design of Chinese knot alike in spirit, which made Chinese knot increasingly popular among people. Today has witnessed a wide range of colorful Chinese knot decorations everywhere. They are used to be as ornament avenues, streets, accessories, present boxes and furnishings.

- 南通民间艺术挂饰 吴元新（现代）
Folk Art Decorations in Nantong, by Wu Yuanxin (Modern Times)

> 各种各样的结

两个绳头缠绕、穿插成一个实体即为结。结之功用为扎系、固定、连接、阻挡，使两绳为一绳，使绳索成环状，使单绳产生阻隔。作为一项古老的技艺，绳结在长期的累积、演进中，呈现出十分丰富的面貌，用途也相当广泛。下面根据结构形态、功能形态和价值内涵的差异对其加以分类阐述。

单结和群结

根据不同的结构形态，绳结大体可归为两大类型：单结和群结。

单结是指那种洗练而完整的单体绳结，它们是最基本的独立结构单位。"单"，即不能再进行分解。平结、云雀结、三环结、十字结等大多数基本结即属这种类型。

> A Variety of Chinese Knots

The knot is defined as passing one end or part of the string through a loop and pulling it tight. The function of the knot is to tie, fasten and connect. As an ancient craft withstanding the test of time, the knot underwent a series of innovation and changes in its structure and function so the design has become all the more diverse. Considering the difference in shape, function and connotation, knots can be broken down into the following categories.

Single Knot and Group Knot

A variety of knots can be fallen into two major categories based on different structures, single knot and group knot.

A complete and undividable piece of knot is referred to single knot, which constitutes an individual unit of group

就功能而言，单结除了其本身的实际功能（如系扣、观赏）外，更重要的是它可以发挥组织、构造的作用，可以成为结的群合形态的起点。就像单字可以组词、造句、写文章一样，单结在重复、延展、群合上具有很大的自由性和适应性。因此，单结编制是掌握绳结技艺的必修课，也是打开绳结艺术世界的一把钥匙。

群结是指通过反复编制一种或数种单结而形成的绳结群。从某种意义上讲，它就是通过绳线或结体彼此连接、重叠、扣合、缠绕而形成的一个结构更加复杂、形体更加庞大的结。"群"，即有可以重复、扩展、联合的意义。就功能而言，线状群结和面状群结更具有广泛的实用性。

knot. Such single knots as square knot, skylark knot, three-loop knot and crossed knot play an important role in connection in addition to their basic functions (tying and decoration). What's more, single knot can act as a joint, just like a sentence and a passage which is composed of individual characters. By duplicating and extending the single knot, group knot will come into being; therefore, single knot is an elementary craft. Without mastering it, you cannot make creative group knots. In a word, single knot helps you get access to the world of artistic knots.

Group knot refers to duplicating one kind or several kinds of single knots. To some extent, a large-sized group knot is formed by overlapping, twining and connecting. Linked by string and knots, group knot can be expanded and duplicated. Actually, group knot is far more complex and functional.

• 群结——三环结的组合
Group Knots, Combined Three-loop Knots

- 单结——纽扣结
  Single Knot, Button-like Knot

- 单结——同心结
  Single Knot, Truelove Knot

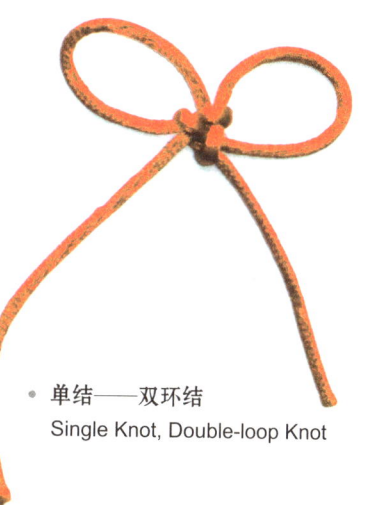

- 单结——三环结
  Single Knot, Three-loop Knot

- 单结——双环结
  Single Knot, Double-loop Knot

- 单结——双钱结
  Single Knot, Double-coin Knot

Culture of Chinese Knot
中国结文化

云雀结
Skylark Knot

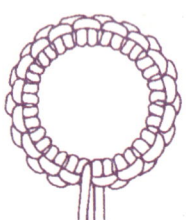

• 群结——云雀结的组合
Group knots, Combined Skylark Knots

块结、带结和网结

绳结在日常生活中都是具有一定功能的。不同的需要造就了绳结不同的形态。概括划分，绳结的功能形态有三大类型——块结、带结、网结。

块结即独立存在、单独承担一定功能的结。相对于"线""面"而言，它的存在形式更像"点"。

Tied Knot, Strap-like Knot and Netting Knot

Knots are used in our daily life. The function of knots varies with the structure. As far as their function and structure concerned, knots can be divided into three types, tied knot, strap-like knot and netting knot.

Compared with the structure of "line" and "plane", tied knot is similar to a

块结是由一根或数根绳线盘绕穿插而成，其种类和编制方法很多，有单结，也有群结。块结依其结构可分为基本结和组合结。基本结即单结，其结构简洁，可单独使用，也可作组合结的基本单位。组合结即点状群结，其结构为多个基本结的组合。组合结有两种类型：一种是重复组合，即以一个基本结为单位反复编制，在编制过程中由于抽拉、连接、重复、交错的方式不同而产生多种形态，如如意结、双喜结；另一种是变化组合，即将几种不同的基本结组合在一起，形成形态变化更加丰富、复杂的结构，如葵花结、金鱼结。

"dot" from appearance because it is self-contained and can be used individually. Either a single string or several strings are knitted in different ways to make knots. Apparently, group knots are formed by multiplying a number of single knots. In general, there are two types of group knots. One is to duplicate a single knot based on different knitting skills such as connecting and interweaving. *Ruyi* knot and double happiness knot are made in this way. The other is to combine several types of single knots, such complicated design and structure as sunflower knot and goldfish knot.

Strap-like knot is to duplicate a certain kind of elementary knot on one string or several strings. Because of the linear extent, it looks like a strap

• 基本结——盘长结
Elementary Knot, *Panchang* Knot

带结是以单根或数根绳线反复编制某种基本结,以一维(线状)的形式依次序不断延伸而形成的,属线状群结。由于它编成后总有一定的宽度或厚度,故称其为"带结"。从形式和用途上看,带结有三种:第一种类似花边,较为稀松,只求形式多样、变化丰富,多作装饰用,如项链、手环、各种边饰。第二种是有一定宽厚度,编制严谨,除要好看之外还要求有较强的耐拉性、柔韧性,如各类腰带、背包带、提带等等。第三种以具有高强度耐拉性、耐磨性为主要目的,通过加粗、扭紧原来的绳线使之更加牢固、坚韧。最典型的是几股绳线编制的"辫结",如三股辫、四股辫、五股辫等。另外还有柱形结,其最为粗实,也最具强

- 基本结——"卍"字结
  Elementary Knot, "卍" Knot

- 基本结——团锦结
  Elementary Knot, Bouquet Knot

remaining a certain width and thickness, hence its name. Considering its feature, the strap-like knot can be used in the following areas. As it is loose and diverse in patterns, it is suitable for edging on necklace and bracelets. Since it is flexible but strong, it can be used as backpack strap and belt. On the other hand, the strap-like knot can be given the greatest wear resistance by adding three-ply, four-ply or even five-ply thread to widen and fasten the original strap. The cylindrical knot is one of the strongest knots. Considering its feature, the strap-like knot is used not only for edging, but also for bundling, tying and hanging objects.

Similar to a net spreading sequentially in two-dimensional form, netting knot is to duplicate a certain

度。带结用途甚广，可作边饰，更多的是作为富有装饰感的绳索，可用来扎、捆、绑、系、挂物品。

网结是以单根或数根绳线反复编制某种基本结，并以二维形式依序平铺开的，属面状群结。它通过一个或数个基本结的重复出现及其余线或环耳的相互连接、延伸，形成一个可以无限发展的面。网结可疏朗空透，也可繁密紧实；可细薄平整，也可粗厚凹凸。由于编制网结的方法不同，可产生千百种不同的花纹变化。网结具有很强的实用功效，成语"天网恢恢，疏而不漏"就是取网结"防范""捕捞"功能的引申义。在日常生活中网结用处甚广，常见的有各类钩花制品（台布、桌垫等），各种编结挂饰（壁挂、帘子等），各种毛衣、线衣，各种罗网、花篮、网袋等等。

kind of elementary knot on one string or several strings, hence its name. If the duplicated elementary knots are connected with loops, this net can extend into infinity. The design of netting knots varies with different knitting techniques. They can be dense, thin, smooth, rough or uneven. The word "net" is of an old saying derivation. So goes an old saying, the net of Heaven has large meshes, but it lets nothing through. Netting knot is widely used in daily life such as crocheted tablecloth, table mat, woven wall hanging, curtains, knitted sweater, baskets and shopping bags.

平结
Square Knot

- 带结——平结的组合
Strap-like Knot, Combined Square Knots

盘长结
Panchang Knot

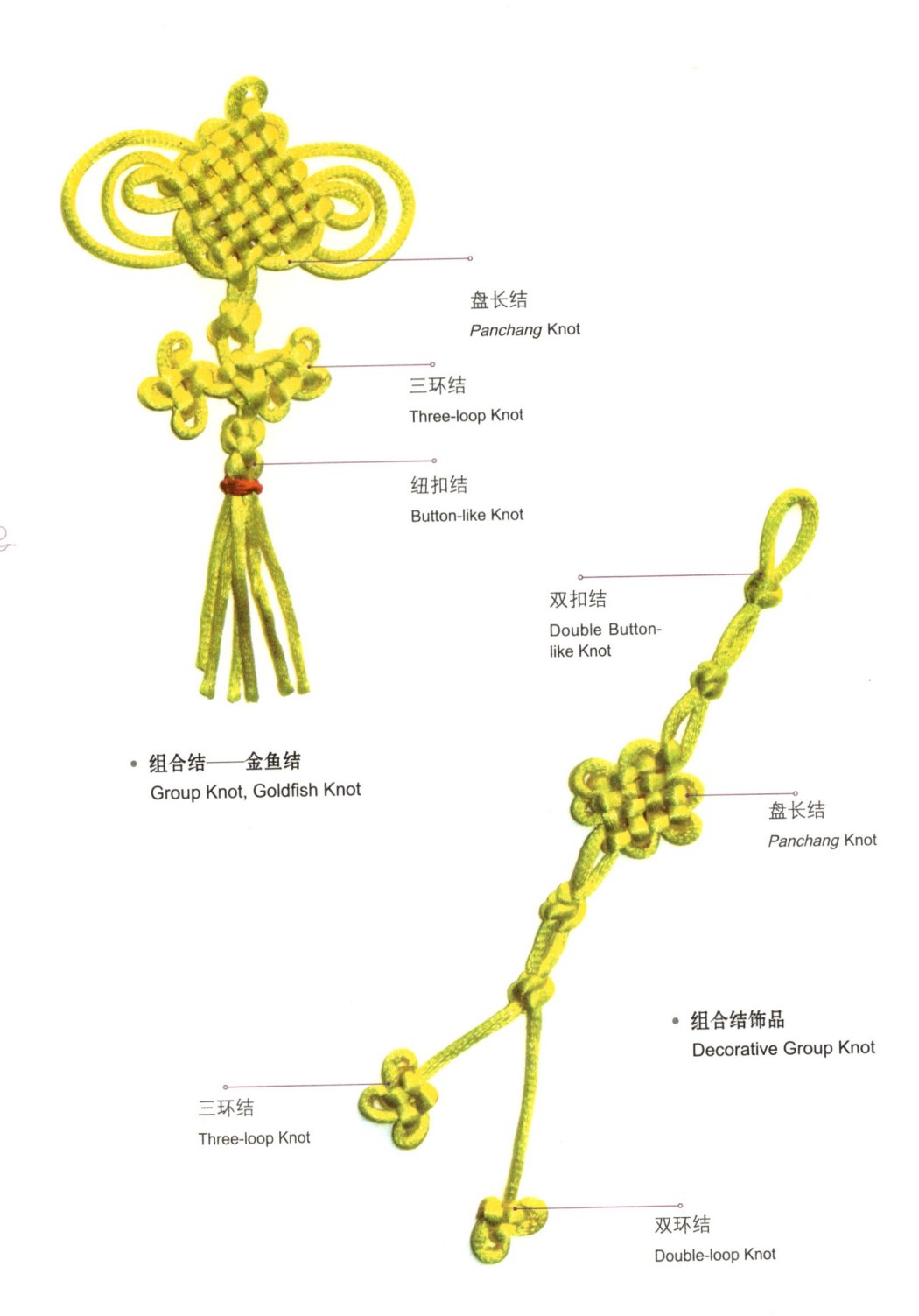

三环结
Three-loop Knot

纽扣结
Button-like Knot

双扣结
Double Button-like Knot

盘长结
Panchang Knot

- 组合结——金鱼结
  Group Knot, Goldfish Knot

三环结
Three-loop Knot

- 组合结饰品
  Decorative Group Knot

双环结
Double-loop Knot

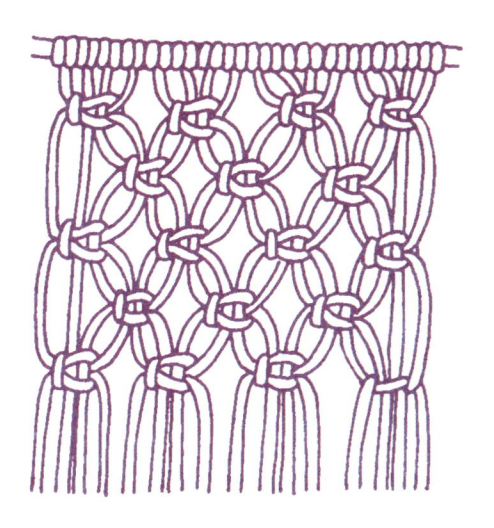

- 网结——平结的组合
  Netting Knot, a Group of Square Knots

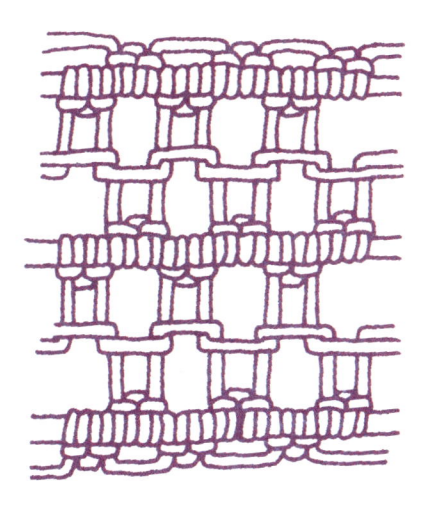

- 网结——云雀结的组合
  Netting Knot, a Group of Skylark Knots

实用结和花结

如果从价值内涵来看，绳结大体可分为实用结和花结（即装饰结）两种类型。

实用结即体现实用价值，起扎系、固定、连接、阻挡等作用的结，如鞋扣、渔网、提襻结等。当然，生活中的实用结，也往往有审美讲究。

花结是反映人们精神需要，体现审美、寓意、象征等精神价值的绳结。它的发展呈两个趋势，一是趋向纯观赏，二是趋向点缀、陪

Functional Knot and Decorative Knot

If you see intrinsic value of the knot, it can be roughly divided into two types, functional knots and decorative knots.

Functional knots play an important role in daily life. Knots are used for tying, fastening, connecting and blocking. As a result, a large number of daily necessities such as buckles, fishnet, etc., are tied with knots. Anyway, even though they act as tools in our daily life, aesthetic considerations have never been overlooked.

People tend to endow decorative

• 实用结——陶罐上的纽
Functional Knot: Handle on Clay Jar

衬。纯观赏的花结，通常以独立或主体的形式出现。其编制的目的是展示绳结本身的结构美和精致奇妙的工艺美。这种结多为群结，结体庞大，组合复杂，变化多端，刻意在结的造型中表现设计者的艺术匠心和艺术创造。作为点缀、陪衬的花结，通常与其他物品结合，以从属的形式出现，起烘托、渲染的作用；其编制的目的是使装饰对象更具风采，更有韵味和情趣。这种结构的结体相对小巧，有单结，也有群结，其形、色、质要根据装饰对

knots with spiritual values, symbolization and aesthetic taste. Generally, decorative knots can be fallen into two types. One is for appreciation. This type of knots is often used individually so as to attract people's attention to the beautiful structure and exquisite craftsmanship. Most of them are group knots showing craftsmen's artistic creation in complex combination and various patterns. The other is for a decorative embellishment. As a foil, the knot usually goes with other accessories so that they contrast with each other and make accessories more

象而定。我们在日常生活中见到的花结，大多属于此类。

　　相对于其他工艺饰品来说，花结的原材料极易获得，一般普通的绳线、丝带、塑料管都可以被利用，而且其编制方法也比较容易掌握。再则，花结可适用的地方很多，如鞋、帽、腰带、项链、耳坠、戒指、手镯、胸饰、服装、钥匙串、挂饰、礼品盒等等。另外，花结在实际使用中也有优势，它不怕挤压、碰撞，即使出现皱褶、变形，也极易修整恢复原样。由于花

graceful and charming. Either single knot or group knot can be as a foil but their size is relatively small. Their color, size and texture are varied in decorated objects. The vast majority of knots in our daily life fall into this category.

　　The material of knot is relatively easy to get compared with other artistic handwork. Depending on an ordinary rope, a ribbon or a plastic pipe, knots can be done. What's more, the skill of knitting is easy to master. Because knots have the advantage of being free from collision, deformation and crease, knots are widely used on costumes, belt, necklace,

• 桌旗上的结饰 秦岱华（现代）
Decorative Knot Pattern on Table Flag, by Qin Daihua (Modern Times)

● 热闹扎堆的结饰
Chinese Knots

• 桌旗上的结饰 秦岱华（现代）
Decorative Knot Pattern on Table Flag, by Qin Daihua (Modern Times)

结的上述特点，使得它与人的关系极为密切。尽管花结没有金银那般耀目，没有珠宝那般华贵，但它却有着自己独特的朴素、自然、灵巧之美。在琳琅满目、多姿多彩的各类饰品中，花结就像一枝绚丽的奇葩，一直受到人们的青睐。

earrings, rings, brooches, key chain, gift box, etc. Even though the knot is not as dazzling as gold and sliver ornaments and not as luxurious as jewelry, simple, natural, ingenious, colorful and diverse as it is, the knot has found favor in the eyes of people.

• 实用结——琵琶结
琵琶结可用作旗袍等中式服装的盘扣。
Functional Knot, Pipa-shaped Knot
It is used as a button on Chi-pao and Chinese style costumes.

• 各式花结
Decorative Knots

> 结的寓意

宋代词人张先在《千秋岁》中写道:"天不老,情难绝。心似双丝网,中有千千结。"(译文:天不会老去,爱情也永不会断绝。相思之心犹如双丝网那千千万万个结扣一样,绵长、纠结。)这首词以网结形容有情人离别后的情感缠绵、纠结的心理状态。在古典文学中,经常会用"结"来比喻人们的缠绵情思和复杂心境。

绳结在漫长的演变过程中,被人们赋予了各种情感和生活的希冀。"结"与"吉"谐音,而"吉"有着丰富多彩的内涵,福、禄、寿、喜、财、安康无一不属于"吉"的范畴,在民间的吉祥图案中,结经常表达的是"吉"的意义。因此,绳结这一具有深厚文化

> Implied Meaning of Chinese Knot

The Song-dynasty poet Zhang Xian wrote in *Thousands of Autumn Passing-by*: "Heaven never grows old, nor will love end. My lovesickness is like an interwound net tied with thousands of knots." This poem depicted after lovers bid farewell, they fell prey to lovesickness. Long and tangled knot as a metaphor for people's lingering emotions and complex state of mind is often seen in Chinese classical literature.

In the long process of evolution, knots have been given a variety of connotations. People adopt homophones and symbolic meanings to endow knots with implicative meanings such as happiness, wealth, longevity and good health so knots are often seen on auspicious patterns of folk art. As the essence of traditional Chinese culture,

内涵的民间技艺作为中国传统文化的精髓，流传至今。

汉语词汇中也经常会用到"结"的概念，如：结合、结交、结缘、缔结、团结、结果、结束、结局、总结、了结、结构、结晶等等。这些词大抵有连接、聚合、终了、构造之意，其含义莫不与人们所熟悉的现实生活中的结有着形与意的联系。

将两绳连接成一绳的结自有连接、聚合的意义；一条直绳打了结后意味着顺直的状态到此终止。结本身构造及关系的巧妙多样，也使人产生联想并借以比喻事物各部分的组织、配合及相互联系。

男女之间的婚姻也常以"结"表达，如：结亲、结发、结婚、结合等。"同心结"自古以来就是男女间坚贞爱情的象征，结饰常成为人们表达情感的定情之物。而"结发夫妻"也源于古人洞房花烛之夜，男女双方各取一撮长发相结以誓相守一生的行为。

"knot" with deep cultural connotation has come down the ages.

Chinese character "knot" often appears in the day-to-day language. Such a single character "knot" constitutes some phrases with meaning of linkage, reunion, ending and conformation in Chinese, so "knot" has become the formation of phrases extending its original meaning.

Two ropes are connected implying junction and getting together. A knotted rope leaving with straight lines signifies down to the end. The structure of knot implies organization and cooperation.

Chinese character "knot" is also related with matrimonial matter. Phonetically, some of Chinese phrases with the meaning of "marriage" derive from "knot" or "knitting". Truelove knot as a token of love has been a symbol of unchangeable and permanent love between sweethearts since ancient times. On the night of wedding day, the bride and groom exchanged a fistful knotted hair as a pledge of their marriage lasting forever.

• 盘长结纹饰品（清）
Accessories with *Panchang* Knot Patterns (Qing Dynasty)

● 西安街头的中国结装饰
Chinese Knot Ornament Along the Road in Xi'an, the Capital of Shaanxi Province

各种结的寓意
Implied Meaning of Different Knots

方胜结
Diamond-shaped knot

吉祥保平安
Safe and sound.

双蝶结
Double-butterfly knot

比翼双飞
Fly side by side; go places together as a happy couple.

如意结
Ruyi Knot

万事称心，吉祥如意
Good fortune as one wishes.

团锦结
Bouquet knot

花团锦簇，前程似锦
Bouquets of flowers and piles of brocades imply having a bright future.

祥云结
Propitious cloud knot

吉祥美好
Good Luck.

双喜结
Double happiness knot

婚姻幸福，双喜临门
Good things come in pairs; a happy marriage.

桂花结
Sweet osmanthus knot

高贵清雅，富贵无疆
Noble, elegance and wealth.

盘长结
Panchang knot

回环贯通，永无终止，长寿百岁
Life goes on and lasts an eternity.

平安结
Safe and sound knot

一生如意，岁岁平安
Peace and good luck all year round.

同心结
Truelove knot

恩爱情深，永结同心
Deep love lasts forever.

蝴蝶结
Bow knot

幸福美好
Happiness.

福字结
Fu character knot

福气满堂，吉星高照
Bring someone good luck and success in life.

寿字结
Longevity character knot

人寿年丰，寿比南山
The land yields good harvests and the people enjoy good health.

磐结
Coiled knot

吉庆祥瑞，普天同庆
The whole nation joins in jubilation.

双钱结
Double coin knot

财源广进，财运亨通
The luck for wealth is prosperous.

• 江苏蓝印被面上的结纹
Knot Pattern on Blue-printed Quilt Cover in Jiangsu Province

中国结的应用
Usage of Chinese Knot

在日常生活中，结的运用也呈现着多样性。结本是实用的，但人们还看重它的形态，力图使实用之结具有美感。于是，人们便不断地在绳与绳的互相缠绕和穿插中寻求巧构变化，以造就美的形态，并以美的形态展现自己的智慧和灵巧，抒发自己对美的追求，寄托自己美好的生活理想。随着时代的演变，编结作为一项传统技艺，依然以活泼的生命力丰富和美化着人们的生活。今天，许多人喜爱结所蕴含的这种东方文化的巧妙神韵，并因其浓郁的民族韵味而称之为"中国结"。

In addition to practical function, knots are designed to meet people's demand for aesthetic value, and as a result, people have been constantly seeking for a change in design and shape. Depending on their wisdom and dexterity, Chinese people have woven the pursuit of beauty, good fortune and auspiciousness into various knots; therefore, diverse decorative knots are not limited to practical function. Knitting knots as a traditional skill has been vitalizing with the times. Decorative knots as the epitome of the charm of oriental culture are popular among a great number of people. Because knots are endowed with rich connotation and a hefty dose of ethnic flavor, knots are referred to as "Chinese Knot".

> 服饰之结

结在服饰中的运用可谓历史悠久，古代服饰中的绳结主要是充当绶带、佩玉和纽扣装饰。流传下来的荷包、香囊、玉佩、扇坠、发簪，以及服装上的盘扣等，无不显示了结在中国传统服饰中的应用之久、之广。

中国古代各个朝代的服饰形制有所不同，所以绳结也各有特色。早在周代（前1046—前256），就有男子佩带玉觿的习俗。觿是一种微曲的锥形器，最初是用来解结的锥子。后来，觿逐渐发展成为佩饰，是成人的象征。到了战国时期（前475—前221），佩玉绳结已有较多样式。河南信阳楚墓出土的彩绘木俑上就清晰描绘了佩玉打结的图案。在东晋画家顾恺之所绘的

> Knot on Costumes

It has been a long history since knots were used on costumes. Knots and frogs were often tied with ribbons and jade pendants in old times. Judging from those frogs and knots attached to purses, perfume sachets, jade pendants, fans, hairpins which were passed down from the old times, it is safe to say that knots have been widely used in traditional Chinese costumes and accessories.

Knots varied with different styles of costumes in each dynasty. As early as Zhou Dynasty (1046B.C.-256B.C.), men started to wear *Xi*, a curved cone-shaped tool for untying the knot. Later *Xi* became a symbolic accessory. The wearer showed that he was in adulthood. By the Warring States Period (475B.C.-221B.C.), there have had various knots matching with jade pendants. From the unearthed painted wooden figurines in

《女史箴图》中，仕女的腰带上就系有单层翼的简易蝴蝶结。随着时代的发展，结的种类更加丰富，出现了"卍"字结、团锦结、十字结等，它们成为服饰中必不可少的装饰。后来出现了酢浆草结，因其形状类似酢浆草而得名。酢浆草是一种三叶草本植物，为掌状复叶，状如蝴蝶，因此酢浆草结又称"蝴蝶结"。这一时期，人们还将单结串联起来，组成各式花结用在服饰中。明清时期，结已普遍应用于人们生活中。

绶带

绶带是中国古代用来悬挂印或玉佩的丝带，一般系在腰间。以绶带区分尊卑是中国古代服饰制度的显著特征。绶带有许多种类，包括印绶、双绶、玉环绶等。

印绶是指系缚在印纽上的彩色丝带，其颜色和长度都有具体的规定，用来区分身份和级别。汉代的官印置于腰间的鞶囊（即绶囊）内，印绶佩挂在腰间，垂搭在囊外，或者与官印一同放入囊内。双绶是穿着礼服时佩带的两条丝带，

Chu Tomb, Xinyang, Henan Province, we can see the pattern of knots and jade pendants depicted on these figurines. A simplified bowknot on the waist band of the court maid was depicted on the *Scroll of Lady Officials* by Gu Kaizhi, a top painter of the Eastern Jin Dynasty (317-420). As time went by, more types of knots appeared such as "卍" knot, bouquet knot, crossed knot, etc. They became an essential part of traditional Chinese costumes. In the Song Dynasty (960-1279), a kind of oxalis shaped knot appeared. Oxalis is a kind of clover, palmate compound leaf, similar to the shape of bowknot, hence its name. During this period, single knots were linked up together to form group knots for costume decoration. By the Ming and Qing dynasties, knots have been widely used in people's daily life.

Shoudai

Shoudai referred to ribbons hanging seals and jade pendants tied on the waist. Wearing *Shoudai* to distinguish people's social hierarchy was a typical feature of the institution of ancient Chinese costumes. There were other types of *Shoudai* such as *Yinshou* (seal and its

通常挂在腰部，左右各一条，由印绶演变而来，南北朝以后广为流行。玉环绶是系挂有玉环的丝织带子，亦由印绶演变而来，南北朝以后较为多见，最初作为礼服的配饰。绶带系于腰间，自然要打结用以固定，同时用花结作为装饰。

colored silk ribbon), double *Shou* (two silk ribbons) and jade ring *Shou* (hanging a jade accessory to press down pleats on the dress).

Yinshou referred to colored silk ribbons tied on seal knobs. There were specific regulations for the length and the color of ribbons to distinguish official ranks and social status. In the Han Dynasty, the official seal was put into a pocket tied on the waist with hanging ribbons outside or inside the pocket. Double *Shou* referred to two silk ribbons tied with jade rings. It evolved from *Yinshou*. Before it became popular in the Northern and Southern dynasties, it was initially hung on the waist with a ribbon on each flank respectively to match formal attire. Jade ring *Shou* was the silk ribbon with a jade ring first used as the accessory of the full dress, also evolving from *Yinshou* and popular after the Northern and Southern dynasties. Decorative knots were tied to fasten silk ribbons.

古代佩双绶的男子
An Ancient Man with Double *Shou*

绶带
Shoudai

绶带上的结饰
Knots on Ribbons

《美人图》（清）
The Painting of Beauty (Qing Dynasty)

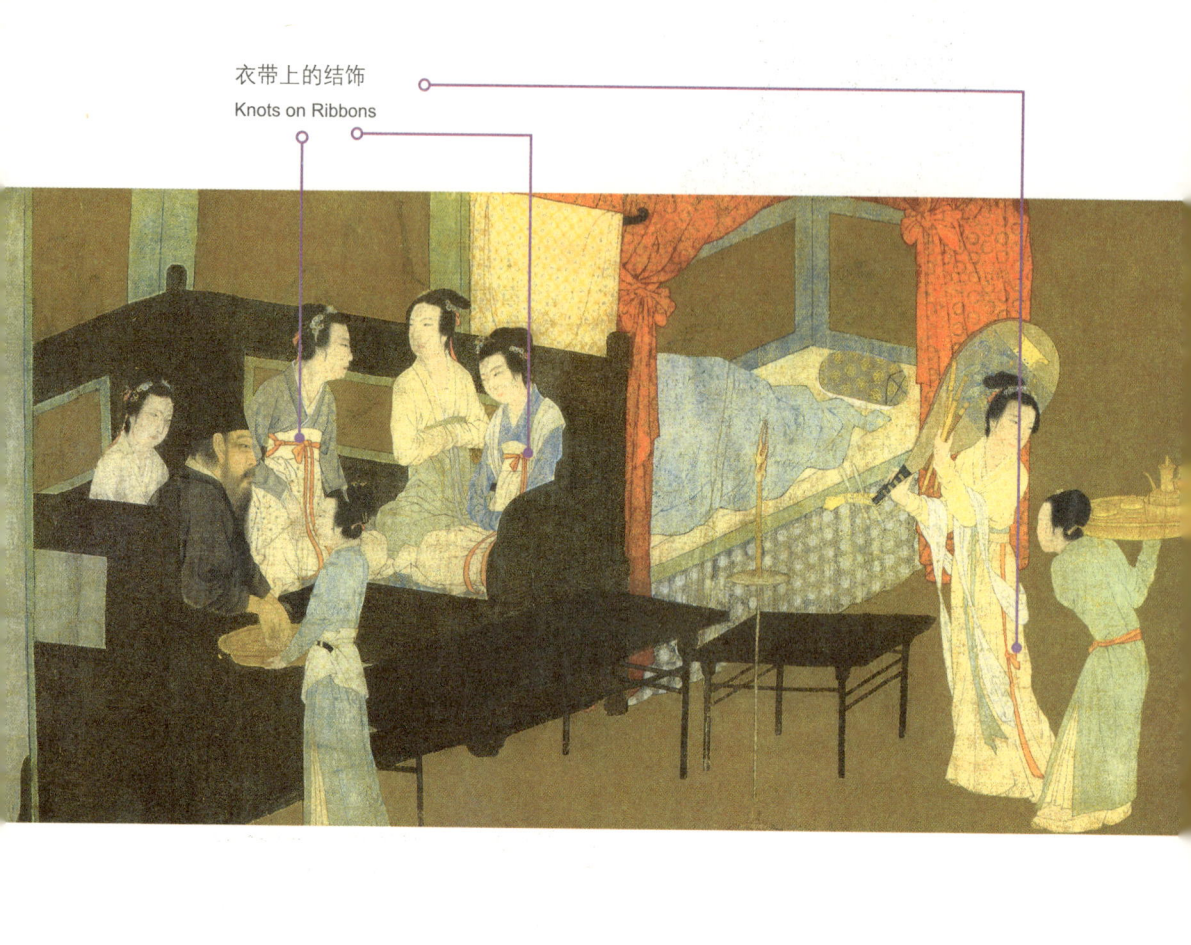

衣带上的结饰
Knots on Ribbons

衣带上的结饰
Knots on Ribbons

- 《韩熙载夜宴图》 顾闳中（五代）
Han Xizai Entertaining Extravagantly, by Gu Hongzhong (Five Dynasties)

- **《千秋绝艳图》中的明代女子**
  画中的女子多数佩有绶带，结饰精致，显出女子的婀娜多姿。
  Ladies of the Ming Dynasty in *The Painting of Court Maids*
  Most of women in this painting wore *Shoudai* tied with exquisite knots, which make them looked graceful and charming.

玉佩

中国人自古便有佩玉的习俗，历代的玉佩大都在其上钻有小圆孔，以便于穿过线绳，系挂在衣服上。

此外还有一种成套的玉佩，称为"组佩"，由好几种不同的玉佩组合而成，而其连接的方法也多靠穿绳打结。玉组佩在商周至两汉时流行，为王公贵族必佩之物。组佩的大小结构根据佩带者地位的高低而不同。汉代以后，玉组佩逐渐消失，到明代时，再度流行，成为冠服制度中不可缺少的佩饰。

古代女子也常系有玉饰，一

Jade Accessories

Chinese people have worn jade accessories since ancient times. In order to link accessories with costumes, most jade accessories had a bored hole to knot the string.

Groups of jade pendants were a combination of multi-piece jade accessories linked by different types of knots. Wearing groups of jade pendants was very popular among princes and aristocrats from the Shang, Zhou to the Eastern Han and Western Han dynasties. The wearers' social status varied with the size and shape of pendants. After the Han Dynasty, the institution related to wearing groups of pendants died away but it resurged in the Ming Dynasty

- 七璜联珠玉组佩（西周）
  Seven Groups of Semi-annular Jade Pendants (Western Zhou Dynasty)

般称为"环佩"。环佩以丝线贯穿，系花结，间以珠玉、宝石、钟铃等，通常系在衣带上，走起路来环佩叮当，悦耳动听。当然，也有人认为腰间的环佩是为了提醒女子不要大步疾行，玉佩叮当作响是不雅的。

盘扣

最早的衣服没有纽扣，所以想要把衣服系牢，就只能借助衣带打结这个方法。后来人们发明了石纽扣、木纽扣、贝壳纽扣，又渐渐发展到用布料制成的盘结纽扣。

盘结纽扣，简称"盘扣"，盛行于清代。当时的服装以袍、褂、衫、裤为主，改宽衣大袖为窄袖筒身，衣襟以纽扣系之，因此盘扣也随着服装的发展而兴起。

盘扣是用缝成细条的布料盘结成的各种各样形状的花式纽扣，其造型优美，做工精巧，是花结的一种。盘扣的花式十分丰富，有拟形的，如兰花扣、梅花扣、金鱼扣、蝴蝶扣、琵琶扣等；有抽象的，如涡旋扣、波形扣等；有盘结成文字的吉字扣、寿字扣、囍字扣等。盘

and became an indispensable part in the institution of costumes and crown.

Ancient women generally wore jade accessories. Their jade pendants were linked up with silk strings tied with decorative knots, beads, gems and bells. They tied pendants on the girdle. When they walked, jade pendants were jingling at every step. Jade pendants were used to remind women of minding their virtuous manners. The wearers had to walk gracefully in case jade pendants would make big sound.

Frog

There was not a single button on original costumes so ancient people had to knot a belt to fasten their clothes. Later, ancient people invented stone buttons, wooden buttons and shell buttons. With time passing by, cloth made buttons and frogs appeared.

Frogs started to prevail in the Qing Dynasty when the fashion changed from loose style into straight clothing with narrow sleeves. Gowns, robes, shirts and pants remained predominantly at that time. The front pieces of costumes were buttoned; the prevalence of frogs accompanied the development of costumes accordingly.

• 一字襟坎肩上的盘扣（北京服装学院服饰博物馆藏品）
Frogs on Straight Breasted Waistcoat (Exhibit in the Costume Museum of Beijing Institute of Fashion Technology)

• 旗袍襟部盘扣（北京服装学院服饰博物馆藏品）
Breast Frogs on Chi-pao (Exhibit in the Costume Museum of Beijing Institute of Fashion Technology)

盘纽的制作
How to Make a Frog

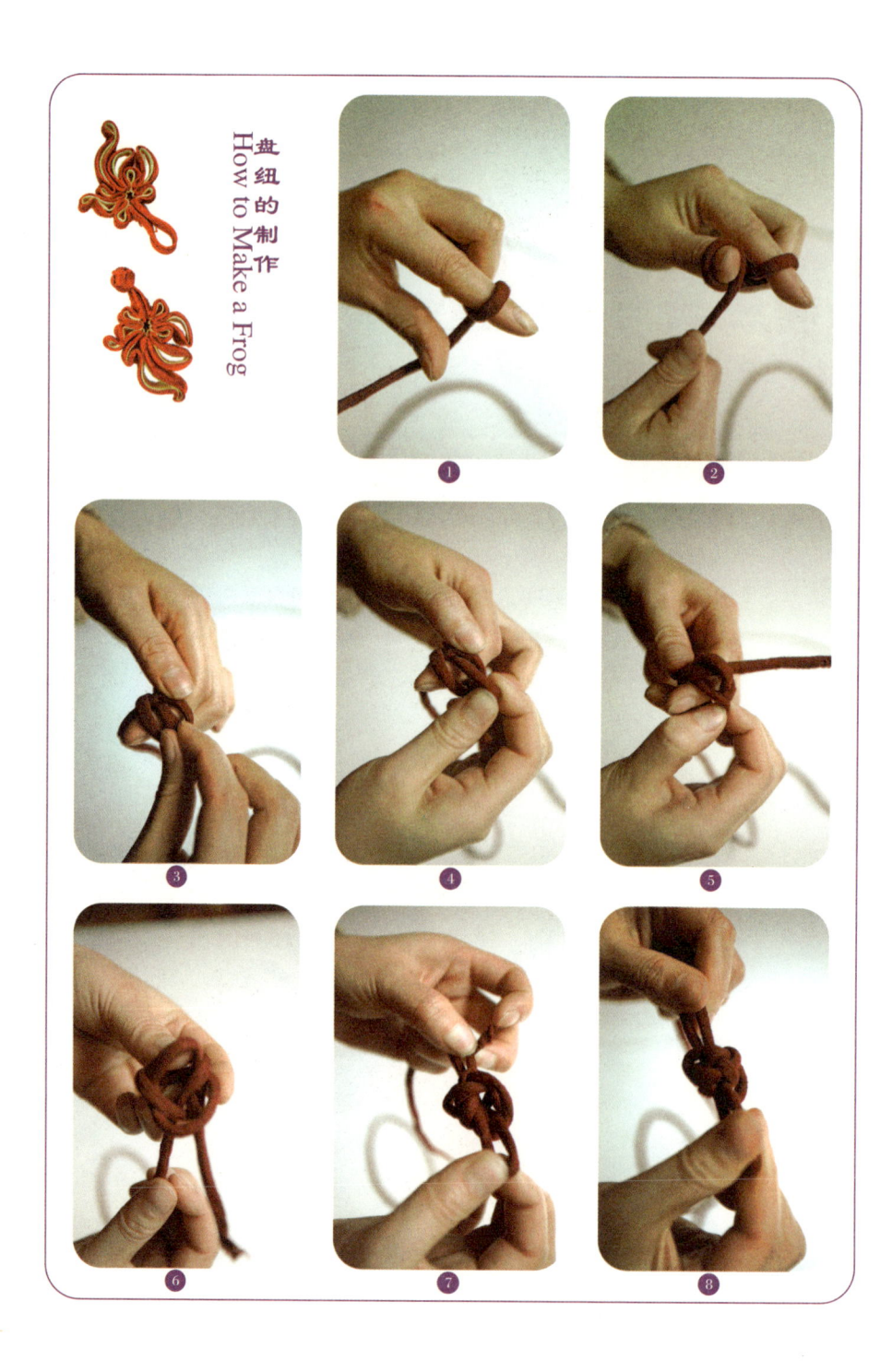

- 旗袍上的盘扣（北京服装学院服饰博物馆藏品）
Frogs on Chi-pao (Exhibit in the Costume Museum of Beijing Institute of Fashion Technology)

扣由组和襻两部分组成，造型既有对称的，也有不对称的，极具实用性和装饰性。其制作工艺包括了盘、包、缝、编等多种手法，在样式设计、颜色搭配等方面也极其讲究。

Frogs made of different strips of cloth look delicate and beautiful. As one of the decorative knots, frogs are diverse in design. From appearance, some are similar to orchid, plum blossom, goldfish, butterfly, *Pipa* (4-stringed Chinese lute), etc. Others are in Chinese character shape, such as "good luck", "longevity" and "double happiness". The frog which is composed of buckle and button in symmetric or asymmetric design is both functional and decorative. Making process of the frog includes winding, wrapping, sewing and knitting. As far as the design and color assortment concerned, they are extremely exquisite.

> 装饰之结

中国结作为装饰，品类十分丰富，用途相当广泛，其装饰对象包括轿子、窗帘、帐钩、笛、箫、香囊、荷包、发簪、烟袋等，且一般有吉祥的寓意。

轿子是一种供人乘坐的旧式交通工具，安装在两根杠上，靠人力抬着移动，有篷或无篷。有篷的轿子，尤其是结婚迎娶新娘的大红花轿，异常华丽，上面多扎红绸并缀花结作为装饰，也寓意吉祥。

过去人们常在窗帘和帐钩上悬挂成串的中国结作为装饰。窗帘的中国结饰通常挂在两侧，形制较大。遮挡蚊虫的帐幔，一般为富贵人家使用。富贵人家的帐幔多为华丽的轻纱，帐钩亦要精致，因此常将中国结或其他吉祥饰物缀于其

> Decorative Knot

A variety of auspicious Chinese knots as a decoration are often seen on sedan chairs, curtains, net hooks, flutes, hair clasps, tobacco pouches and so on.

Sedan chair, a means of transportation, enclosed chair with or with no canopy mounted on horizontally placed parallel poles and carried by men. Bridal sedan was extremely gorgeous decorated with red silk and knots full of auspicious meaning.

People used to hang curtains and net hooks with strings of knots as decoration. Generally, large knots were hung with curtains on both sides of the windows. Hooks were used to hook veiling which prevented insects from entering. Wealthy people chose the superb veiling and exquisite hooks. Chinese knots or other auspicious ornaments were stitched as decoration.

上，不仅起到了装饰的作用，且寓意喜庆和美好。

笛和箫是中国的传统吹管乐器，用竹子或金属制成。古时的好乐者，常以吹奏箫笛为乐，甚至随身携带。人们一般会在笛和箫远离吹口的一端挂上坠饰，或为中国结，或为以绳结串起的玉佩、珠串等。

香囊是一种盛香料的小囊，以绸布或金属制成，佩于身或悬于帐以为饰物。荷包是随身佩带的一种

Xiao and flute made of metal or bamboo are traditional Chinese pipes. In old times, most music lovers took pleasure in playing *Xiao* or flute and had one with the person. People usually hang a Chinese knot or a string of jade pendants and beads on the mouthpiece.

Sachet was a small bag filled up traditional Chinese medicine and spices. It was made of metal or silk. People used it as an ornament and either wore it or hung it on the veiling. In ancient times, pouch was designed to hold such necessities as money, seal and

- 绳结绑的泥哇呜（宁夏民间乐器）
Clay *Wawu* with Braided Ropes (Folk Musical Instrument in Ningxia)

用来装钱币、票据、印章、首饰等零碎物品的小包，另外还有专放烟丝、烟叶的烟荷包及放针线的针线荷包。香囊与荷包都兼有实用功能及装饰功能，还是旧时青年男女的定情信物。女子常将自己平时佩带的香囊或亲手绣制的荷包赠予男方，以表达自己的爱慕与牵挂。在绣制香囊或荷包时，人们常坠以长长的花结，不仅为了使其更秀丽、漂亮，也寓意百年好合、吉祥如意。

发簪是女子用来固定和装饰头发的一种首饰。以前人们常在发簪上装缀一个花枝，并在花枝上垂以珠玉、花结等饰物，称为"步摇簪"。

烟袋是吸水烟或旱烟的用具，由烟袋锅、烟袋竿和烟袋嘴构成。烟袋竿通常系有坠儿，可以是丝线打成的花结，或是玉饰、珠串等。

除了物品的坠饰，中国结还可作为陈设品单独悬挂，或制成各种形状的小工艺品。

jewelry. In addition, there were special pouches holding pipe tobacco, needle and thread. In addition to functional and decorative role, they were seen as a token of love in old times. A girl often gave her embroidered sachet to her beloved expressing her affections. When making sachets and pouches, women never forgot to stitch a long decorative knot to beautify their needlework implying the meaning of "a harmonious union lasting a hundred years" and "good luck".

Women wore hairpins to ornament and fasten buns. They usually decorated floral shaped accessories with jade pendants and knots hanging on hairpins. This kind of hairpin swayed with women's each step, hence its name

Tobacco pouches acted as a tool for water pipes and long-stemmed Chinese pipes. The whole piece was composed by the bowl of the pipe, tobacco holder and the mouthpiece of the pipe. There were pendants, beads, jade accessories and knots hanging on the pipe holder.

In addition to matching with other accessories, Chinese knots can be made into different-shaped small handicrafts used as furnishing.

• 花轿
Bridal Sedan Chair

有结饰的箫

Xiao with a Decorative Knot

• 《吹箫图》唐寅（明）
The Painting of Playing Xiao, by Tang Yin (Ming Dynasty)

- 宫廷香囊及烟荷包（清）
  Imperial Perfume Satchels and Tobacco Pouch (Qing Dynasty)

• 有结饰的荷包（清）
Pouches in Shape with Knots (Qing Dynasty)

• 清代皇妃钿子上的绳结纹
Rope Knot Pattern on Decorative Crown of Imperial Concubine (Qing Dynasty)

- 荷包上的结饰（近代）
  Knots on Pouches (Modern Times)

- 香囊上的结饰 年碧华（现代）
  Decorative Knots on Perfume Satchel, by Nian Bihua (Modern Times)

• 甘南卓尼禅定寺门饰
Door Decoration of Chanding Temple in Zhuoni County, Gannan Autonomous Prefecture in Gansu Province

> 礼仪节庆之结

中国结作为中国传统文化的符号，常被用于礼仪节庆活动中，如七夕节、春节、端午节或婚嫁等喜庆的日子。

七夕节

每年农历的七月初七是中国的传统节日"七夕"，又被称为"乞巧节""女儿节"。七夕节是古代未出嫁的女子们最为重视的节日之一。这一天，姑娘们会手绑彩线，乞求可以像天上的织女一样心灵手巧。七夕节的传统活动就是乞巧，乞巧的方式一般是穿针引线，制作一些小物品，比赛谁家女子的女红精巧。

唐代诗人林杰曾写过一首描写民间七夕乞巧盛况的诗《乞巧》：

> Chinese Knot for Festivities

Chinese knot as a symbol of China's traditional culture is often used for ceremonial festivals and activities such as Double Seventh Festival, the Spring Festival, Loong Boat Festival and the wedding day.

Double Seventh Festival

Double Seventh Festival also known as *Qiqiao* Festival is on the seventh day of the seventh lunar month. It was an important day for girls in ancient times. On that day, they would make colored ribbons in the expectation of being as quick-witted and nimble-fingered as Weaver Maid. The traditional activities on that day were mostly related to needlework competition, such as sewing and knitting some small items.

"七夕今宵看碧霄，牵牛织女渡河桥。家家乞巧望秋月，穿尽红丝几万条。"（译文：七夕节夜晚仰望天空，似乎看到了牛郎和织女在喜鹊桥相会。家家户户在月下乞巧，穿梭织绣的红丝线何止几万条啊。）

各地的七夕节各有特色，乞巧活动的内容也各不相同。例如广州的乞巧节，姑娘们会预先准备好彩纸、线绳等材料，编制成各种奇巧的小玩意，其中少不了花结，还要将谷种和绿豆放入小盒里用水浸泡，使之发芽。节日当晚，姑娘们要穿上新衣服，戴上新首饰，焚香燃烛。七夕之后，姑娘们将自制的小工艺品互相赠送，以示友情。在浙江的杭州、宁波等地，这一天姑娘们会用面粉制作各种小物件，用油煎炸后称"巧果"。夜晚，她们便在庭院内陈列巧果、莲蓬、白藕、红菱等，然后将事先捕来的蜘蛛放在盒子里。如果第二天开盒时，蜘蛛在盒子里结了网便称为"得巧"，是个好兆头。

As the grand occasion depicted in the poem *Qiqiao* by Lin Jie, a poet in the Tang Dynasty, on the night of Double Seventh Festival, looking up to the sky, Cowhand and Weaver Maid were dating on the bridge of magpies. Each household was busy with embroidery and knitting to pray for them. Tens of thousands red strings were woven into cloth.

Activities held on Double Seventh Festival varied in different areas. In Guangzhou, girls prepared colored paper, string and other materials beforehand for knitting a variety of exquisite gadgets on that day. In addition to decorative knots, green bean seed and seed-corn were soaked into water to make germination. Girls wore new clothes and accessories burning incense. After the festival, girls would exchange homemade handicraft for their friendship. In some areas of Hangzhou and Ningbo in Zhejiang Province, girls used flour to make various buns. After deep fry, they were known as "Double Seventh Festival Fruit". They put "fruit", lotus seeds, lotus roots and red caltrop in the courtyard, and then put spiders in the box. If the web was done by spiders the next day, it would be seen as a good omen.

• 有结饰的挂饰（清）
Decoration with a Knot (Qing Dynasty)

067
Usage of Chinese Knot
中国结的应用

春节

春节是中国的农历新年，从腊月（农历十二月）二十三到正月（农历一月）十五，以除夕（农历十二月三十）和正月初一为高潮，是中国人最重要的传统节日。春节期间，中国人会举行各种各样的活动以示庆祝，如祭奠祖先、辞旧迎新、迎禧接福、祈求丰年等。

在古代，每逢春节除夕，长辈就会用红绳系上百枚铜钱作为压岁钱，称为"长命百岁结"。近年来，人们把用红绳编织成的盘长结、福字结、如意结、双鱼结等各式花结，用作春节期间室内悬挂的装饰物，或是互相赠送的礼物。其优美的造型、古色古香的韵味给传统佳节增添了喜庆、吉祥的气氛。

Spring Festival

The Spring Festival is the most important traditional Chinese festival, known as China's Lunar New Year, from the twenty third of the twelfth lunar month to the fifteen of the first lunar month. New Year's Eve and New Year's Day are the climax of festivities. A variety of ceremonies and activities will be held to celebrate the Spring Festival such as holding a memorial ceremony for ancestors, bidding farewell to the old year and praying for good harvest in coming year.

In ancient times, the elder tied a hundred copper coins with red string on New Year's Eve, known as "longevity knot". In recent years, people weave various knots such as *Panchag* knot, *Ruyi* knot, happiness character knot, double-fish knot, etc., to decorate the room or present to each other. These beautiful, ageless and auspicious knots add color to the traditional festival.

- 春节时的结饰
  Decorative Knots in the Spring Festival

- 春节门厅装饰
  Home Decoration in the Spring Festival

中国结的颜色寓意

中国结的颜色非常丰富，应用起来也是极有讲究的，尤其是在节庆日，大都使用红色的中国结。中国人自古以来就崇尚红色，因为古人认为红色是一种吉祥色，象征喜庆、热情、正气、进取。中国人的一生都离不开红色：传说情侣是月老给拴上红绳才结成了良缘；结婚时新娘从上到下、从里到外要穿红色，还要盖上红盖头，坐大红花轿；女人生孩子，要请亲朋好友吃染成红色的鸡蛋；过年时老人要给小辈发"红包"，家家户户门上要贴红色的对联和"福"字；年轻人立功受奖时要戴大红花，接受红色的证书……所以，红色的中国结是节庆时的首选。

《红楼梦》中就有一段为玉佩打络子（结）的情节描写："宝钗笑道：'这有什么趣儿，倒不如打个络子把玉络上呢。'一句话提醒了宝玉，便拍手笑道：'倒是姐姐说得是，我就忘了。只是配个什么颜色才好？'宝钗道：'若用杂色断然是不好的，大红的又犯了色，黄的又不起眼，黑的又过暗。等我想个法儿：把那金线拿来，配着黑珠儿线，一根一根地拈上，打成络子，这

• 中国结饰品 金媛善（现代）
Decorative Knot, by Jin Yuanshan
(Modern Times)

才好看。'"这一段描写的是为一块白玉配编一套花结。在这段引文中，"大红的又犯了色"指的是红色与白色相冲（五行中的金以白色为代表，火以红色为代表，火克金）；而一块质地洁白的玉，"用杂色断然是不好的"；黄色与白色搭配明度近似，故而"不起眼"，起不到互相衬托的效果；而黑色与白色对比又过于强烈，所以"过暗"。五行中的水以黑色为代表，金也可以金色为代表，金生水，所以文中宝钗提议的以金线配着黑珠儿线则十分符合中国的五行配色理论，寓意吉祥，颜色也非常古朴谐调，可见古人编络子的艺术修养。

The Implication of the Color of Chinese Knot

Chinese people are particular about the choice of colors in different festivities. Because red has been regarded as an auspicious color symbolizing enthusiasm, righteousness, the spirit of enterprise, bright future, Chinese people have a preference for red, especially in festivities. Red has been an essential color to accompany Chinese people in their lives. As a legend goes, the old man under the moon unites persons in marriage by fastening the red string between lovers. Sitting in a red sedan chair, the bride had to wear red all over and put on a red veil. After giving birth, women had to treated

- 中国结饰品
  Decorative Knot

families and friends to dyed red eggs. During the Spring Festival, the elder gave "red envelop" to kids. Red couplets and "happiness character" were pasted on the doors of each household. After the young were rewarded for a meritorious deed, he would have the honor to wear a large red flower. At the same time, officials would send a red bulletin of glad tidings. In a word, Chinese people prefer red Chinese knot in festivities.

There was a depiction about the color of knitting knots in *Dream of the Red Mansion*, a masterpiece of the Qing Dynasty. "You may tie the jade with knots as well," Baochai told Baoyu with a smile. "Of course, cousin." Baoyu clapped his hands in approval. "I'd forgotten that. But what color would be best?" "Nothing too nondescript would do," said Baochai, "But crimson would clash, yellow wouldn't stand out well enough, and black would be too drab. I suggest you get some golden thread and plait it with black-beaded thread to make a net. That would look handsome." This depiction was about a white jade pendant matching with the knot. Baochai took advantage of the theory related to Five Elements (Metal, wood, water, fire and earth, held by the ancients to compose the physical universe. According to the theory of Five Elements, red, black, white and gold represent fire, water and metal respectively, among which fire restricts metal and metal promotes water.) to match colors. In addition, it was endowed with an auspicious implication. It is obvious that people in ancient times had high level of aesthetic knowledge.

• 石雕上的结饰
Knot Pattern on Stone Carvings

• 花结饰品
Decorative Knot

端午节

端午节是中国的重要传统节日，为每年的农历五月初五，民间活动十分丰富，有划龙舟、挂艾草、吃粽子、饮雄黄酒等。

端午节时，家长还会给孩子穿"五毒衣"，系"百索"，以保佑孩子平安健康。五毒衣是具有驱毒虫寓意的衣服，以杏黄色的布缝制而成，上面绣有蝎子、蜈蚣、蛇、蜘蛛、蟾蜍的图案。

● 端午节街头卖香囊的老妇人（图片提供：徽图）
The Old Vendor Selling Sachets on Loong Boat Festival

Loong Boat Festival

Loong Boat Festival, an important traditional Chinese festival, is on the day of May 5th in Chinese lunar calendar when there are a variety of folk activities such as loong boat race, hanging Asiatic wormwood, eating *zongzi* (traditional Chinese rice-pudding) and drinking realgar wine.

On the Loong Boat Festival parents had their children wear "clothing with the pattern of five poisonous creatures" and "plaited silk thread" to bless them safe and sound. Five poisonous creatures are scorpion, viper, centipede, house lizard and toad, whose patterns were embroidered on a piece of apricot yellow cloth implying being away from poisonous creatures.

"Plaited silk thread" also known as "colorful plait" and "longevity plait", was braided with five colored silk threads, red, yellow, blue, white and white. On the Loong Boat Festival, they were knotted or twisted before people hung them on the door, bed curtains, cradle, etc. Also, they were tied on children's wrist and neck. "Plaited silk thread" had following five types : wearing twined five colored thread on children's wrist;

百索又称"长命缕""五彩缕",是以红、黄、蓝、白、黑五种颜色的丝线编结的绳子。端午节这天,大人会以五色丝编成结或拧成绳,悬挂在门上、床帐、摇篮等处,或戴在孩子的项颈及手臂上。百索的形制大致有五种:将五色丝线合股成绳,系于孩子的手臂处;在五彩绳上缀饰金属饰物,挂在孩子的项颈上;将五彩绳编成方胜结,挂于孩子的胸前;将五彩绳编

putting the metal ornaments on five colored thread and then wearing them on children's neck; knitting five colored thread into diamond-shaped knots and suspending on children's chest; weaving five colored thread into the figure of a man and suspending on children's chest; presenting the five-colored embroidered design of the sun, the moon, and stars or the design of birds and beasts to the elder.

Nowadays some people knitted five colored *zongzi*-shaped accessories as

• 粽形结（图片提供：FOTOE）
Zongzi-shaped Ornament

制成人像，戴在孩子的胸前；将五彩丝线绣绘成日月星辰、鸟兽等的形状，敬献给长辈。

如今还有人在端午节时编制五彩小粽形饰物戴在身上作为吉祥物。这种五彩粽子的内壳是用硬纸叠成的，外面编有五彩丝线。

a talisman. Cardboard was made into *zongzi*-shaped model and five colored silk thread twines around it.

端午节佩带香囊的习俗

在端午节，妇女和小孩会佩带五彩的香囊，香囊里面装的多是香料与草药。佩带香囊有很多讲究，小孩一般佩带虎、豹子等动物形香囊；成年人佩带象征着夫妻恩爱、家庭和睦、万事如意的图案或形状的香囊，如菊花、梅花、桃子、苹果、荷花、娃娃抱公鸡、双莲并蒂等；年轻人常常将香囊作为传达情意的礼物，绣上鸳鸯、比翼鸟等图案，由女方于端午节前送给男方。

The Custom of Wearing Sachets during the Loong Boat Festival

During the Loong Boat Festival, women and children wear the sachets filled with all kinds of traditional Chinese medicine and spices. People are particular about wearing sachets; children have beast-shaped sachet such as tiger, leopard, etc.; adult wear sachets implying conjugal affection and family harmony such patterns as lotus, plum blossom, chrysanthemum, peaches, apples, baby holding a rooster and double lotus. The sachet is seen as a token of love. The girl often gives her embroidered sachet with the pattern of mandarin ducks and inseparable king birds to her beloved expressing her affections.

- 虎镇五毒纹兜肚（清）

  古时，"虎镇五毒"兜肚多在端午节期间穿着，多为儿童所用。图中这件兜肚将五毒的形象拟人化，头部及上身以五位身着彩衣的女子代替，下半身仍是毒虫形象。

  Doudu with the Pattern of Five Poisonous Creatures (Qing Dynasty)

  In ancient times, people had their children wear "*Doudu*" (a kind of underwear) with the pattern of five poisonous creatures. They were personified. Their upper body and head were replaced by five ladies wearing colored clothes while the lower body remained as it was.

传统婚礼

从古至今，"结"始终充当着男女相思相恋的信物，将一缕缕的丝绳编制成结，赠予对方，表达了忠贞不渝的爱情和绵绵的思恋。在中国传统婚礼上，盘长结和同心结是必不可少的装饰，寓意一对新婚夫妻永远相依相随。

Traditional Wedding

Chinese knot has been acting as a token of love since ancient times because it was made of strings of silk thread symbolizing unbroken love and unswervingly loyal to the beloved one. *Panchang* knot and truelove knot were essential decorations in Chinese traditional wedding symbolizing a newly married couple would go hand in hand with eternal love.

- 传统婚礼的模拟场景（图片提供：FOTOE）
A Simulated Scene of Traditional Chinese Wedding

附录：几种中国结的编制技法
Appendix: Knitting Techniques of Several Types of Chinese Knots

　　中国结的编制需要经过编、抽、修等程序。每种结都有自己的编制"线路"，编法是固定的，但是"抽"可以决定结体的松紧、耳翼的长短、线条的流畅与工整。只有熟记这些线路，掌握编制要领，多加练习，才能编好中国结。本部分根据结构形态和组织方式，分基本结和组合结两部分，介绍结的编制技法。

Each Chinese knot has its fixed knitting route. Knitting, pulling and trimming are essential process but tightness of the knot, the length of the loop and the neat and orderly outline can be adjusted by pulling the thread. Only by memorizing the knitting routes, mastering the techniques and practicing more, can we knit the knot skillfully and exquisitely. Based on the shape and structure of Chinese knots, they are divided into elementary knot and group knot as introduced in this book.

> 基本结

　　基本结就是常见的、较为简单的独体花结（即前文所述的"单结"），它们本身可以作为小的装饰，但更多的则是作为变化丰富的"大结"的结构基础或基本单元。对于初学者来说，应首先掌握基本结构的编法，然后举一反三，不断地加以扩展、变化。在熟练掌握之后，再根据自己的构思和兴趣，进行创新。

> Elementary Knot

As one of the decorative knots, the structure of elementary knot tends to be simple and independent (as stated above "single knot"), which constitutes the elementary part of complex knots. It is very important for a green hand to master the knitting technique of elementary knots because you can draw inference about more complex knots from elementary ones. After mastering the technique skillfully, you will be able to innovate and develop the design, based on your preference and conception.

结的编制工具和材料

编制花结虽不算太难，但要编得快、编得好、编得巧，就要注意方法和技巧。特别是一些较为复杂的、花色较多的"大结"，更要注意在材料和步骤要领上多加考虑。

工具

编制花结其实不需要什么特别的工具，在普通劳动妇女那里，做个结儿、扎个穗子全凭一双灵巧的手。但为了使初学者方便，这里介绍几种辅助工具，以便在编制较复杂的结时不至于把线路搞乱。

（1）图钉、大头针、纸盒盖或包装用的泡沫板：主要用于固定绳线，搞清编制线路的来龙去脉，便于绳线的盘绕、穿插。

（2）镊子、钩针：当初学者用图钉、大头针将绳线按线路固定在纸盒盖或泡沫

- 编结的工具
  Tools for Knitting Knots

板上，并用手指捏着线头在复杂的线路网上穿插时，经常会觉得手指太粗、碍事，这时可利用镊子、钩针等工具，但镊子、钩针均不宜太尖，以免拉毛或拉坏绳线。

（3）竹签、针、线、剪刀：结编好后，还需抽紧、修整。有些地方还要借助竹签来挑、拨、送；有些需挂坠的吃力之处或易松散之处，要用针线钉缝一下，以固定结形。当然，女红中的重要工具——剪刀也是必备的，用以修绳剪线。

材料

编制花结的材料以绳线为主，其种类十分丰富，诸如以棉、毛、丝、麻、尼龙、混纺、塑料、皮革等材料制成的绳、线、带、条，甚至纸绳、草绳、金属线等。通常只要能够盘绕、抽拉，有一定柔韧性的各种线材，都可以用于编制花结。

就初学者而言，绳线的软硬度最好适中。若太硬，穿插、抽拉就不便，而且结形不易控制；若太软，花结轮廓就不清晰，结形易松垮不精神。当然，对熟练者来讲，绳线的软硬各有利弊，只要利用得好，各种材料都能形成其独特的风格。

一般情况下，编制花结用的绳线应单纯些，如果绳线本身的纹路、色彩过于繁杂，编成结后会让人眼花缭乱。而且，如果结的纹与色和绳的纹与色互相干扰，那么，结所特有的那种缠绕之美、纹路之美、层次之美也会丧失。

另外，材料中还包括花结上常用的珠、环、管、坠子、纽扣等饰物，这些材料用得恰当的话，可以通过反衬、点缀的方式增加结的美感。

Tools and Material of Knitting Chinese Knot

Although knitting knots is not too complicated, it requires paying special attention to knitting techniques and methods if you want to do it well, quickly, and skillfully. Special attention should be given to material and steps when making the complex and varied "large knot".

Tools

There is no need to prepare special tools. Depending on a clever hand, ordinary women can make a knot and tassels. Several auxiliary tools are introduced to make each step clear enough for beginners.

(1) Thumbtack, pin, foam board or strawboard: they are used to fix the thread and make the entire process clear and wind the thread easily.

(2) Tweezers and crochet hook: when doing the first step mentioned above, beginners may find their fingers are all thumbs so tweezers and crochet hooks are helpful to carry the thread; however, tweezers and crochet hooks should not be too sharp, otherwise they will burr the thread.

(3) Bamboo stick, needle, thread and scissors: when knitting is done, using these tools to pick

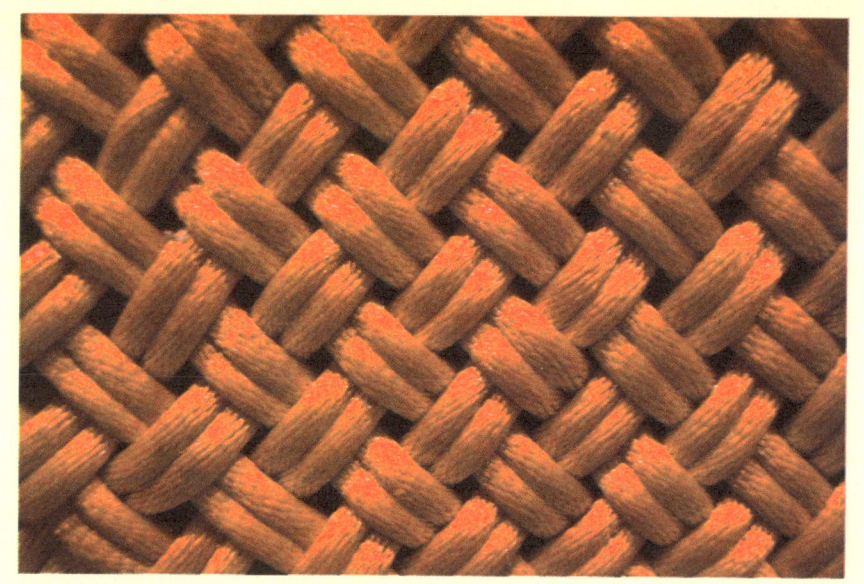

- 盘长结细部
Detailed Picture of *Panchang* Knot

up and move the thread, tighten and adjust knots is necessary. To fasten the structure of knots, needlework is essential to stitch the pendant and loose areas. As the necessity of needlework, a pair of scissors is needed to trim the thread.

Material

The main material of knitting knot is thread and rope. As long as the texture of thread is flexible enough suitable for winding, coiling and pulling, it can be used as knitting material. Such thread and rope are made of cotton, wool, silk, linen, nylon, blending, plastic, leather, paper, straw and wire.

The rope used by beginners is neither too stiff nor too soft otherwise the woven knot will be out of shape easily. The knot will become too stiff to interweave and pull up or too loose to shape. If you are skillful enough, you can take advantage of different materials so as to form a unique style.

Generally speaking, the thread used for knitting knots tends to be simple. If the thread is diverse in texture and color, they will outshine the knot in winding structure and dazzling design. A collision with the pattern of knots could ruin the beauty of knots.

In addition, knots are often embellished with such material as beads, rings, tubes, pendants and buttons. Used in proper, they serve as a foil to knots.

结的编制要领

工具、材料准备好后，就可以确定结形开始编制了。一般程序是编初形、整形、定型。编制花结时除了注意编制顺序外，还要注意以下几点：

1. 保持绳线平整

编制花结所使用的线材有圆的、扁的、方的、双色的等等，编制时应该注意绳线的平整、顺直，不要使之扭曲、反拧，这是初学者常易忽视的。另外，如用双线或多线编制花结，为避免繁乱或线与线的纠缠，应先用单线或最重要的那根线编制初形，编完以后不要抽紧，再将第二根、第三根线依次按原线路顺第一根线穿入，穿时同样要注意线本身的顺直和线与线之间的整齐，这样才能保证结的平整、美观。如果花结中要配珠、环或坠子之类饰物，千万别忘了有些是在编的过程中就应该穿好的。

2. 抽拉均匀

编制花结需心静，切不可急于求成。依步骤将结的雏形编完后，有顺序、按方向地将其抽紧、抽匀是很重要的，这直接关系到结形的美观。抽的方法、方向以及程度不同，都会形成不同的结形。一般来说，要先将结心抽紧，然后慢慢调整外沿和环耳部分。在调整时，要看清线的来龙去脉，注意绳线的平顺，由一端（起始处或终结处）向另一端慢慢调节，此时切不可胡乱拉扯，否则将乱作一团，前功尽弃，必要时还可借助竹签或镊子等工具。

3. 仔细收尾

花结编完调好后，就该进行最后的收尾处理了。收尾的任务是完善每一个细部，如以针线钉缝需要加固、修整之处，镶缝各类珠饰，巧妙处理线头（或藏于结内，或打一小结作为装饰，或穿入珠、管、或连接成环状），等等。此阶段更需耐心、细心，力求精致。注意钉、缝、镶、藏都应尽量不露针脚，不露痕迹，给人以自然而成之感。结的魅力之一在于精巧，仔细收尾是达到精巧必不可少的一步。

Key Points of Braiding the Knots

When the material and tools are ready, we can start knitting knots. General procedures are as follows: setting up an initial shape, shaping and finalizing the design. In the process of knitting order has to be followed. Also, the following points should be noted.

1. Flattening the thread

Keep the thread flat and straight. Do not warp the thread. Some beginners often overlook this point when knitting knots with bicolored thread, round-shaped or flat-shaped or square-shaped threads, etc. Weaving bicolored knot or multi-colored knot has to determine a single thread as a basis to set up an initial shape so as not to tangle together. Do not pull the thread to make the knot taut before the second thread or the third thread following suit. In the process, keep each thread neat and straight so as to make sure the finished knot pleasing to the eye. If beads, rings or pendants are necessary, they should be strung in the process.

2. Pulling the thread evenly

Knots cannot be done in a rush and knitting requires patience. Following the instruction to finish the initial shape, and then pull the thread tautly and orderly. It is a key step because it will determine the appearance of the knot. The style of the knot varies with the method, direction and tightness of pulling the thread. In general, tighten the center of the knot prior to adjusting the outer loops. Be sure to pull the thread from one end to the other step by step. See to it that the thread remains neat and straight. If necessary, such tools as bamboo sticks and tweezers can be used. More haste, less speed. Do not pull the thread at random otherwise the artwork will be messed up.

3. Winding up knots

The final process is sure to improve each detail after adjusting the structure of knots. Detailed perfection includes fixing all seams, inlaying decorative beads and hiding thrums delicately (using the remaining thread to knit a knot as decoration; inserting thrums into beads, tubes or linking them up to be in loop shape). Patience is a must to do this process. Depending on the process of sewing, inlaying and hiding, all thrums will be removed so that knots show a sense of natural beauty. Exquisiteness is the soul of knots so the last process cannot be too ingenious.

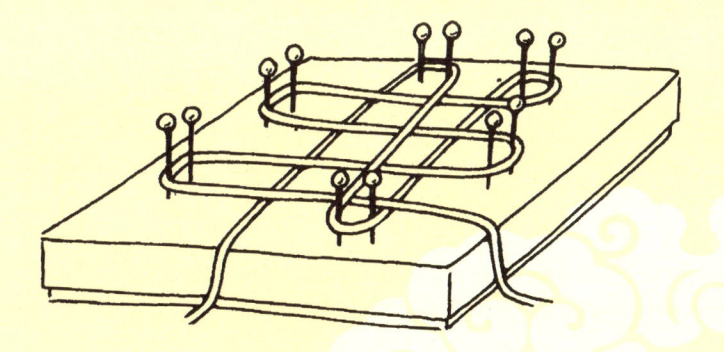

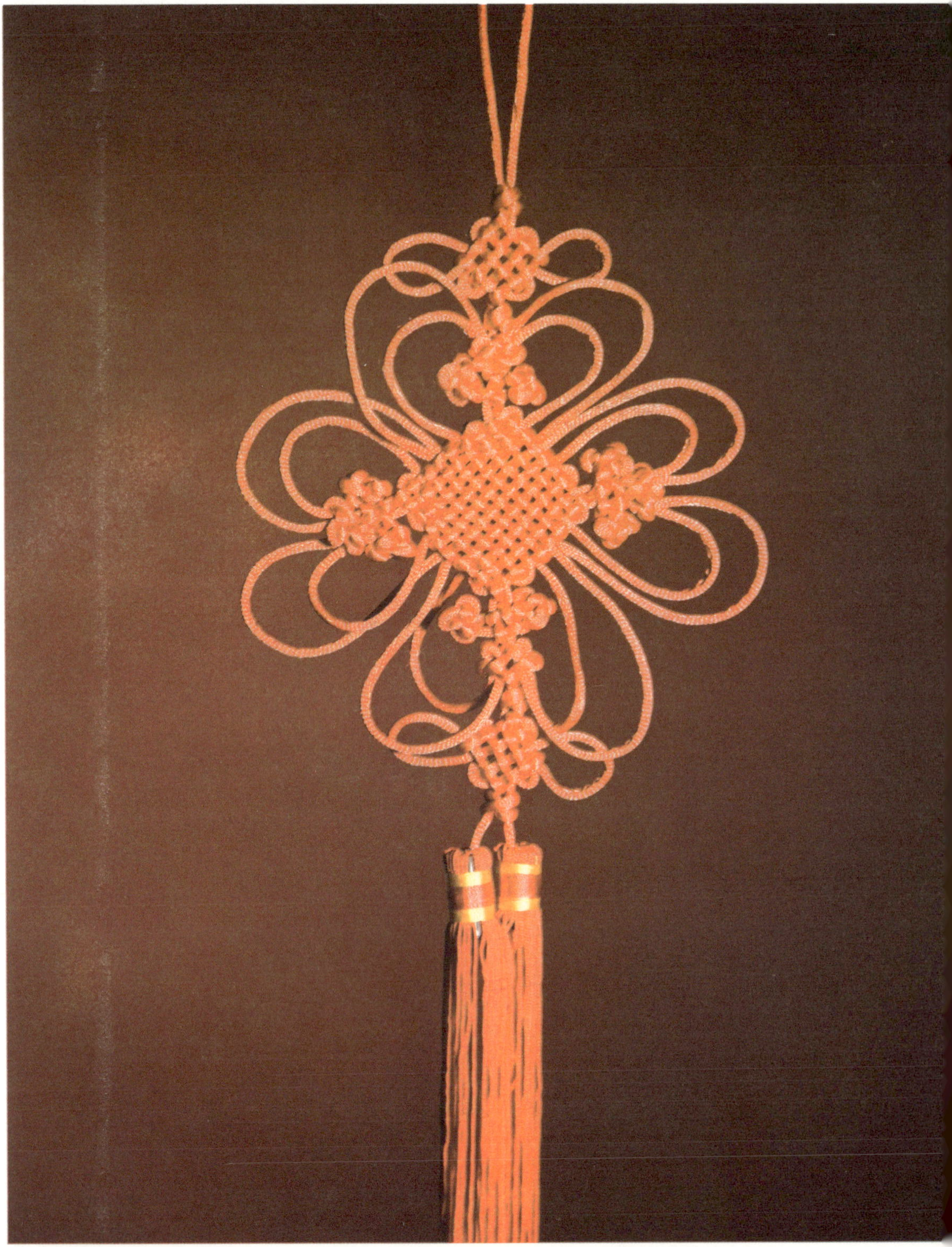

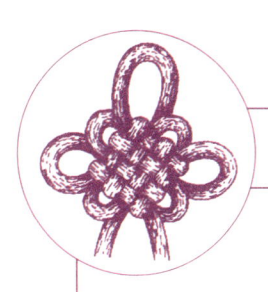

盘长结
Panchang Knot

  盘长结的形状颇似佛家"八宝"之一"盘长",故而得名。编制盘长结时切记线路走向。结编完后,将两个剩余线头相接,大有不分始终、循环无穷之感。此外,由于盘长结有八个环耳,所以又被称为"八结"。"结"与"吉"读音相近,民间常以"结"喻"吉",故盘长结又称作"八吉"或"八吉祥"。盘长结的结构致密、结体结实、结形优美,除常单独使用外,还有许多变化或搭配,可演变出十分丰富的花结来。

The shape of *Panchang* knot is similar to one of the Buddhist eight treasures, *Panchang*, hence its name. Knitting *Panchang* knot has a settled weaving route. The remaining thread on each end is connected together giving a sense of cycle. Because there are eight loops on *Panchang* knot, it is also known as "eight knots". As "knot" is the phonogram of "auspiciousness" in Chinese, *Panchang* knot is also endowed with auspicious meanings. The structure of *Panchang* knot is close-grained and beautiful. A variety of decorative knots are derived from *Panchang* knot so it can either match with other ornaments or use individually.

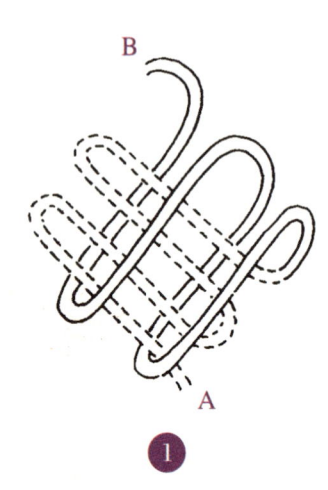

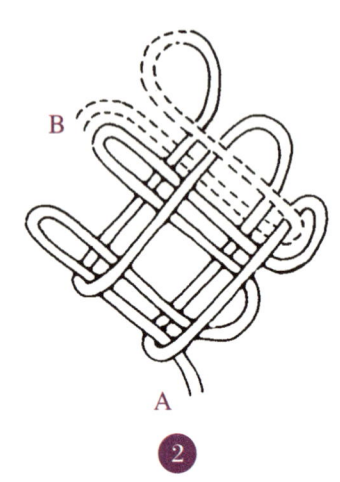

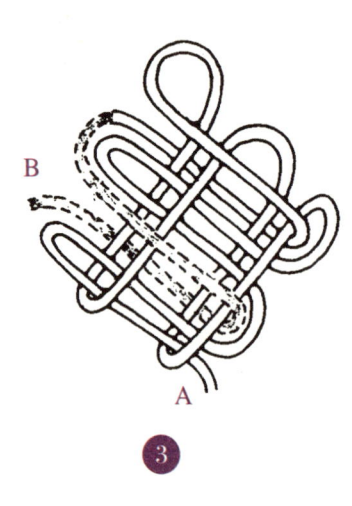

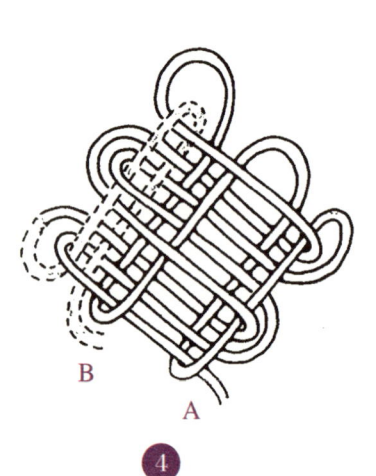

制作步骤
Knitting Process

① ② ③ ④

B A　　　　　A
　　　　　　　B

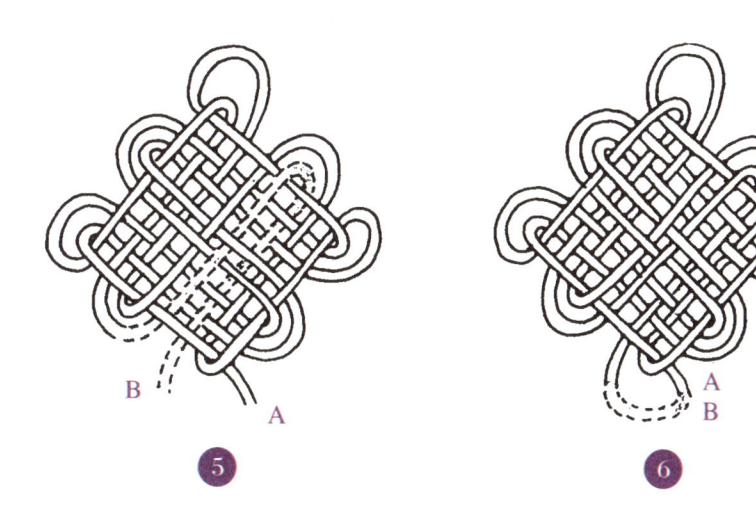

⑤　　　　　⑥

⑦

089

附录：几种中国结的编制技法

Appendix: Knitting Techniques of Several Types of Chinese Knots

同心结
Truelove Knot

  同心结是一种古老而寓意深长的中国结。由于其两结相连的特点，常被用来象征男女间的爱情，取"永结同心"之意。另外，若将同心结中间的两绳向外拉，又可衍化出美丽的"卍"字结。"卍"字结的结心图案很像"卍"，由此得名。

As an old and meaningful Chinese knot, truelove knot implies love between men and women because of its inseparable design. If two threads in the center of truelove knot are pulled outwards, it will change into a beautiful "卍" pattern.

制作步骤
Knitting Process

方法一
Method 1

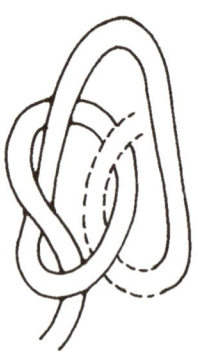

① ② ③

制作步骤
Knitting Process

方法二
Method 2

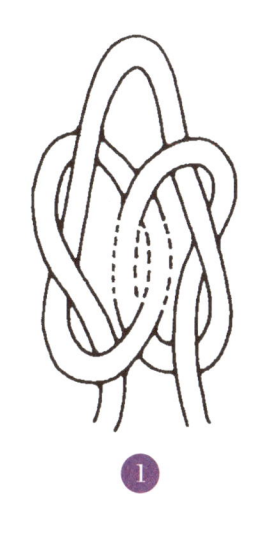

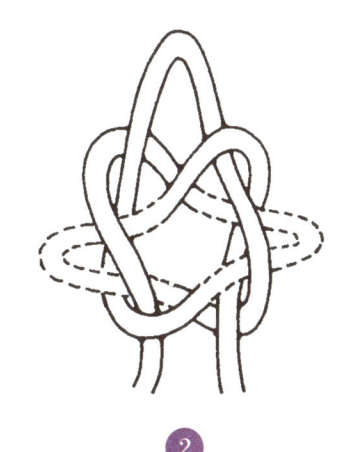

① ② ③ ④

091

附录：几种中国结的编制技法
Appendix: Knitting Techniques of Several Types of Chinese Knots

纽扣结
Button-like Knot

　　纽扣结是人们最熟悉的一种中国结，常用作服装的扣子，也称"葡萄扣""钻石结"。其编法甚多，结形也不尽相同。这种结不易松散，且结心可藏线头，所以常作收尾结。纽扣结编完后的修整十分重要，应将绳线自顶部向下慢慢按顺序抽拉，注意保留顶部的线圈高度，切勿抽乱。另外，如不将此结抽紧，而作拉平处理，则成为另一种扁平的花结，也很有装饰性。

Button-like knot is familiar to us because it is often used as a button on clothes, also known as "grape button" or "diamond knot". The shape of knots varies with different knitting techniques. This type of knot is tight in structure so it is often used to wind up the thread neatly. All the threads have to be slowly pulled from the top in proper order seeing to the height of winding on the top. If the winding is flattened out instead of remaining loose, it will change into another decorative knot in oblate shape.

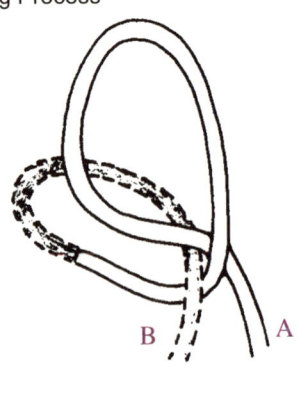

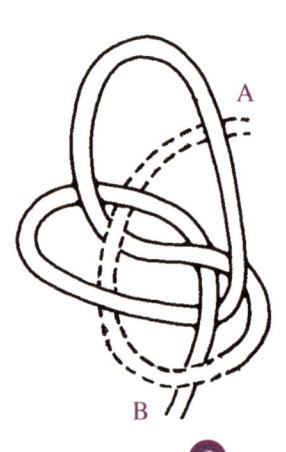

制作步骤
Knitting Process

093

Appendix: Knitting Techniques of Several Types of Chinese Knots

附录：几种中国结的编制技法

三环结
Three-loop Knot

三环结的用途甚广，因为它是用一个线头进行穿插编制的，所以在编任何一个带环耳的花结时，都可在上面作一个三环结，以增加变化。用类似三环结的编制方法，还可以编出双环、四环、五环等一系列花结。

It is widely used to vary the pattern because in the process of weaving any kinds of knots with a loop can be connected with three-loop knots. By duplicating loops, knots can be woven into double-loop knot, four-loop knot, five-loop knot and so on.

制作步骤
Knitting Process

方法一
Method 1

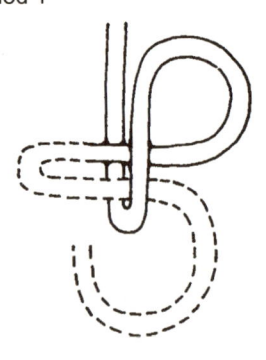

① ②

③ ④

制作步骤
Knitting Process

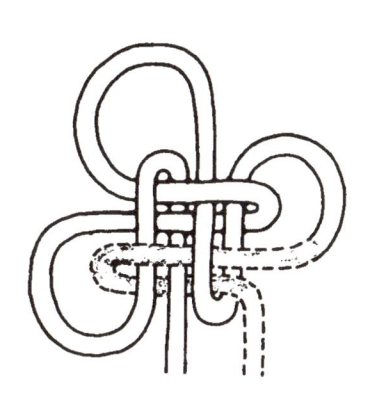

方法二
Method 2

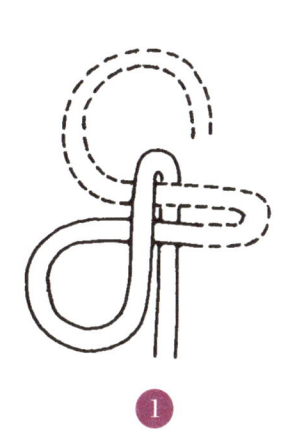

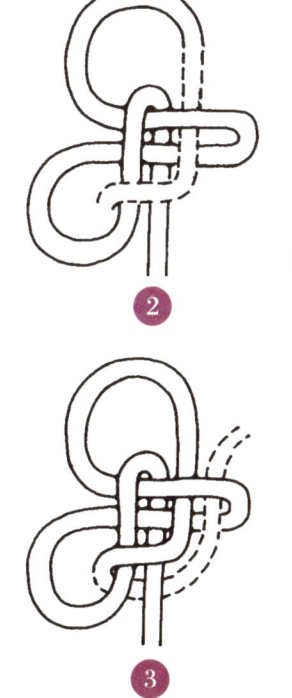

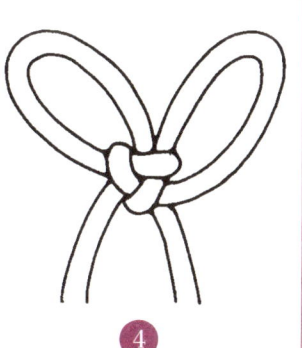

① ② ③ ④

草花结
Club Knot

草花结是一种流传久远、甚为常见的中国结。其结形朴素、美丽且多变，就像田野里不知名的花草一样。草花结编法十分简单，起头时做几个耳，编好后就有几个"花瓣"。编时可单层一次编成（结的两面图案各异），也可重复再编一层（结的两面图案相同），使其形状更加丰厚美丽。

Club knot is a common design handed down through many generations. It is simple but varied and beautiful just like the unknown grass and flowers in the field. The number of "petals" on the knot depends on the quantity of loops staring from knitting. Weaving one layer will show different patterns on both sides while duplicating this layer will appear the exactly same pattern on both sides. Either way can enrich the pattern. Overlying the patterns will thicken and beautify the knot.

制作步骤
Knitting Process

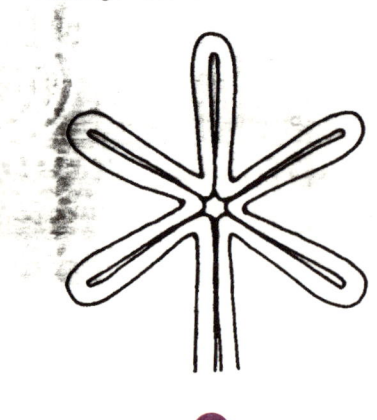

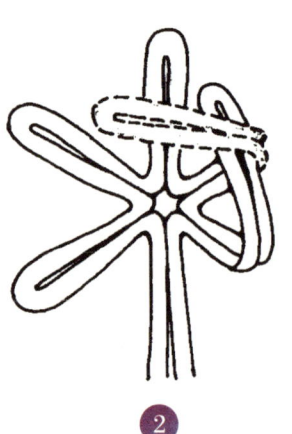

❶ ❷

③ ④ ⑤

⑥ ⑦

097

附录：几种中国结的编制技法

Appendix Knitting Techniques of Several Types of Chinese Knots

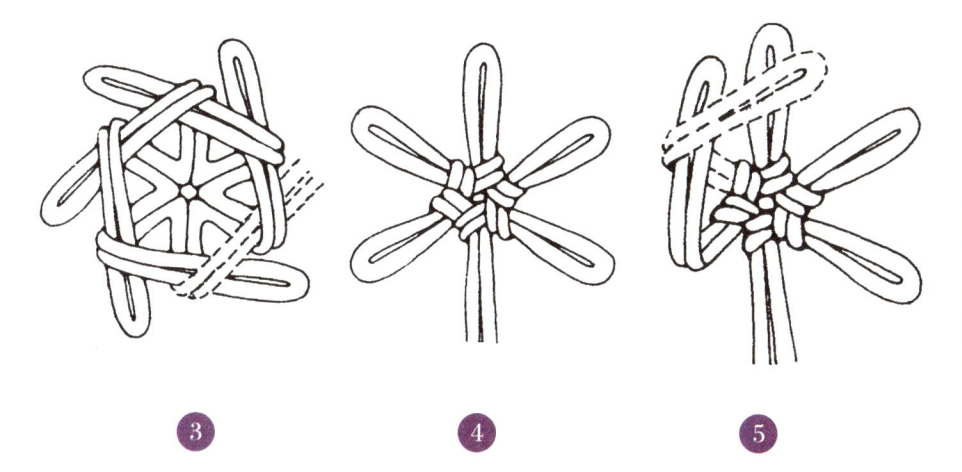

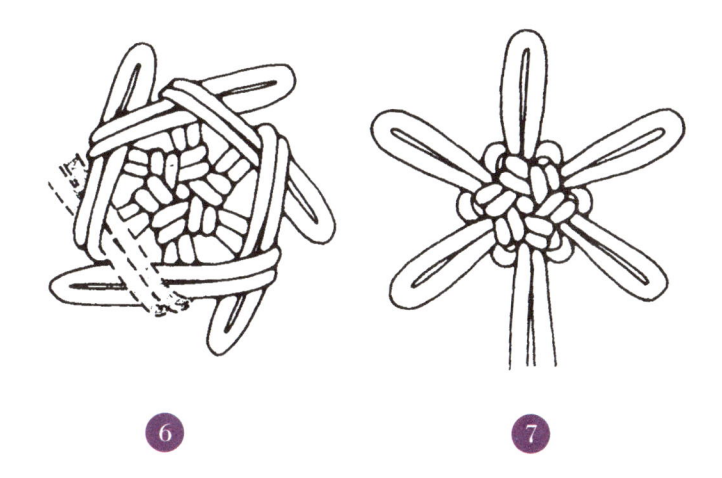

双钱结
Double-coin Knot

　　双钱结又称"金钱结""双元宝",因结形很像两个叠在一起的铜钱而得名。制作时可从绳线的一头来编,也可两头同时编。此结拉紧时平整美观,若保持一定松度,有意留些空隙,则更似"双钱",另有一番装饰趣味。

Double-coin knot also known as "money shaped knot" and "*Yuanbao* (a shoe-shaped gold ingot) shaped knot", looks like two stacked coins, hence its name. It is woven either from one end of the thread or on both ends as long as keeping the proper space and looseness when the knot is flattened. Interspace will make "double-coin knot" ornate and highly decorative.

制作步骤
Knitting Process

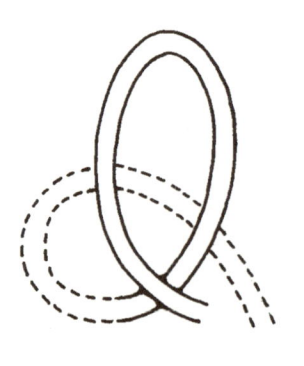

①　②

③　　　　　　　　　　④

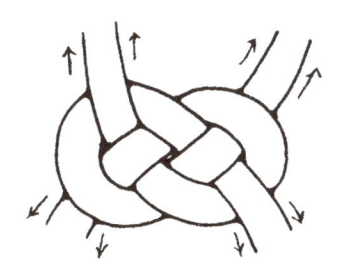

双钱结的组合应用
Combined Double-coin Knots

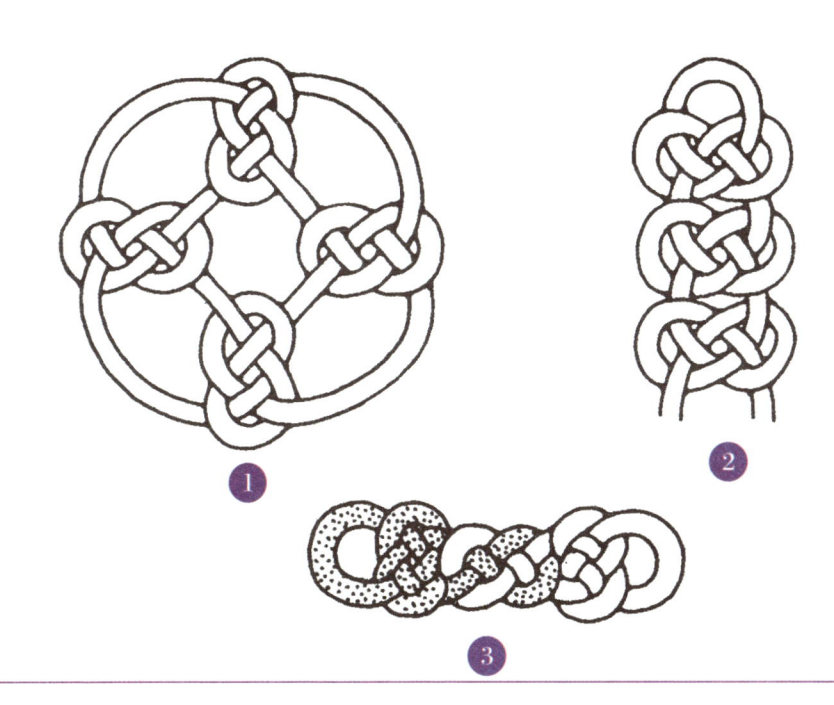

①　　　　　　　　　　②

③

绶带结
Knot Tied on *Shoudai*

绶带结为古时绶带上常见的结饰，是以单线双股编成。其结心正面、反面不同，适合正面观赏。抽拉调整时要注意绳线的平顺，将双重环耳调成外大内小。

Knot tied on *Shoudai* knitted by two strands of a single string was very common in ancient times. The center of the knot as a front view looked better than its reverse side. The size of the inner loops was adjusted to be smaller than the outer ones when the complete knot was straightened.

制作步骤
Knitting Process

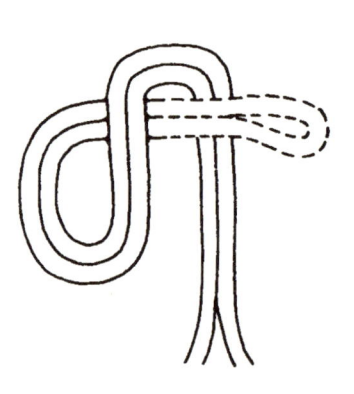

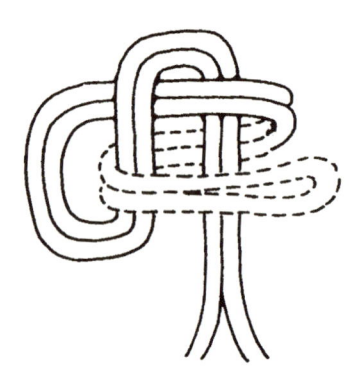

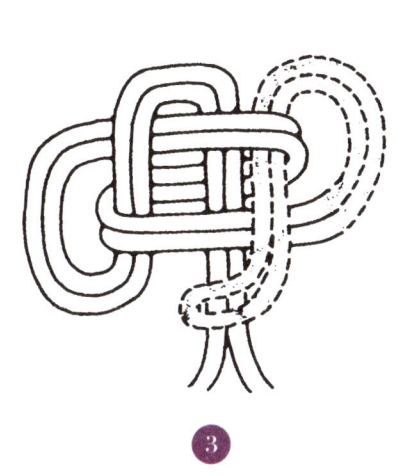

③

④

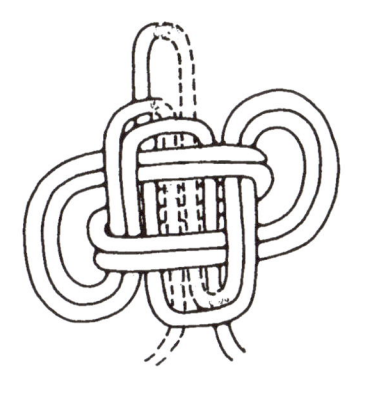

⑤

⑥

101

附录：几种中国结的编制技法

Appendix: Knitting Techniques of Several Types of Chinese Knots

团锦结
Bouquet Knot

团绵结是一种多瓣、圆形的中国结，中心有孔，可镶小珠。

・将绳线依中点分A、B两端。

・当A端自右向左走时，从所有遇到的线下穿过；自左向右走时，从所有遇到的线上压过。

・B端沿自身方向逆行向上，然后自左向右时，遇到自身线或接触自身线的线从底下穿过，遇到其他线则从上面压过。

・当B端钩着右面环耳回到左边时，从各线下面穿过，依原线路回到起点。

It is a kind of round multi-petal knot with a hole in the center to inlay beads.

• A and B on each end of the thread with reference to the midpoint.

• Once threads encounter with A wind beneath when A is woven from the right to the left. All the threads are wound above when A is woven from the left to the right.

• B goes upward in opposite direction, going from the left to right, winding beneath all threads once encountering B; encountering other threads, winding above the rest of threads.

• When bringing the loop from the right to the left, B is to wind beneath all threads back to the starting point.

制作步骤
Knitting Process

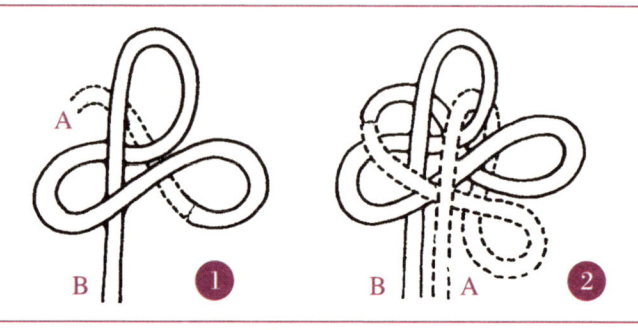

103

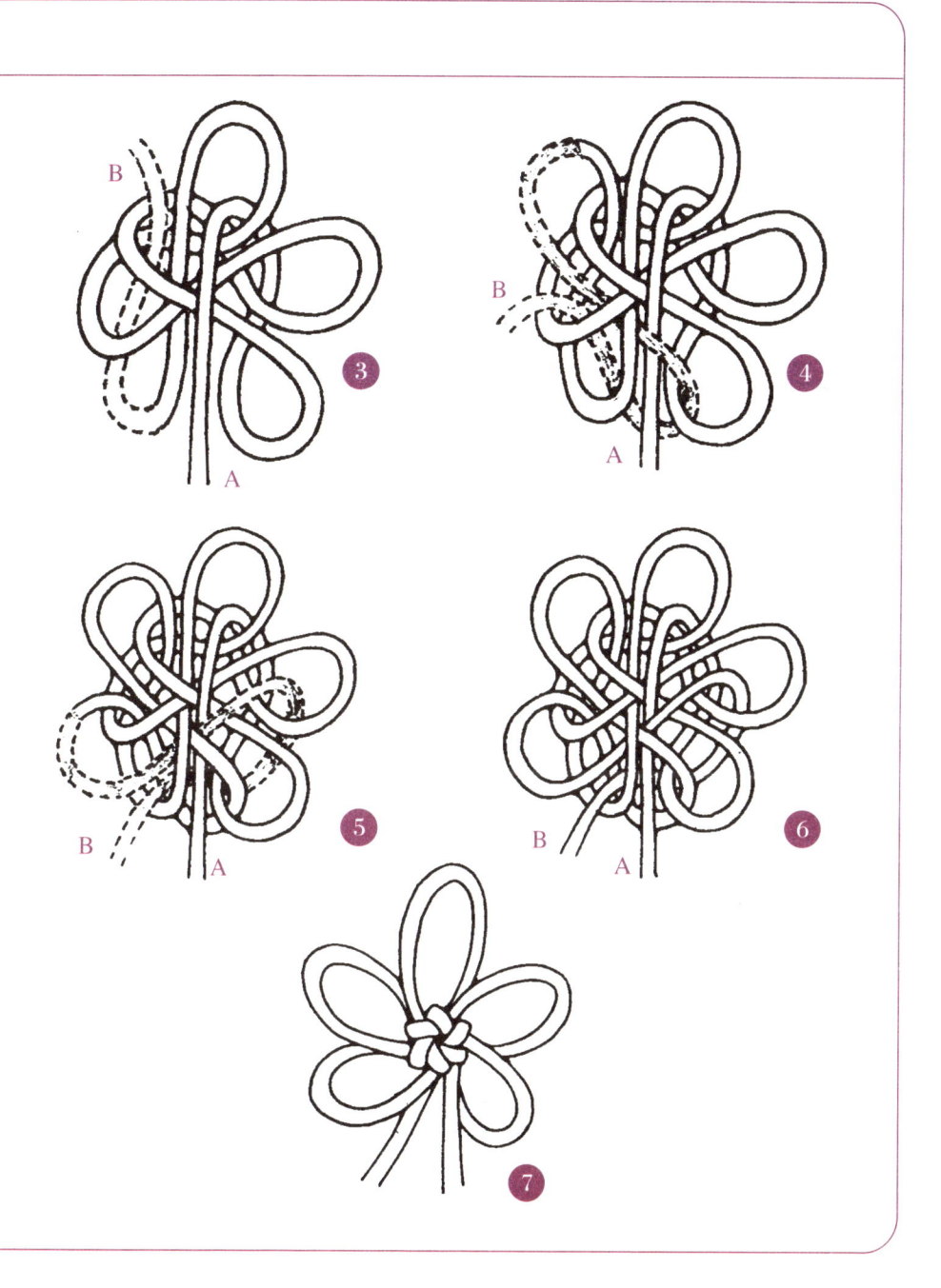

附录：几种中国结的编制技法

Appendix: Knitting Techniques of Several Types of Chinese Knots

> 组合结

　　组合结是由几个基本结组合而成的结。它是一种多体花结（即前文所述的"群结"）。其组合方式有两类。一类是重复组合，即以一个基本结为单位反复编制，在编制过程中由于抽拉、连接、重复、交错的方式不同而产生多种形态变化；另一类是变化组合，即将几种不同的基本结组合在一起，其形态更加丰富，也更具装饰性。基本结具有单纯之美，但在许多场合下，会显得单薄、孤立，如果将它们作多种组合、搭配，就会显得丰富且具有变化，其使用范围也会广泛得多。

　　掌握了基本结的编法后，即可进行组合结的设计和编制了。在设计组合结时，首先要考虑结的用途、结的外形，然后根据用途和外形确定结的材料、结构和编法。一

> Group Knot

　　Group knot, also known as combined knot, is composed of several elementary knots. It can be divided into two categories. One is to duplicate an elementary knot. The structure of group knot varies with the method of pulling, connecting and interleaving. The other is to combine different types of elementary knots, which diversifies the decorative group knot. Although elementary knot looks simple in shape, it is pleasing to the eye. Compared with elementary knot, group knot tends to be diverse so it is widely used.

　　Group knot will be easy to make as long as the knitting technique of the elementary knot is mastered. The usage and shape have to be given consideration before you decide on the material, structure of the knot and knitting technique accordingly. In general, people

般组合结都要以一个主体为中心。它既是视觉的焦点，又是连接、延续、缀挂其他结饰的主体。在编制时，不但要注意结形本身的美观，还要注意结与结之间的疏密关系、色彩关系及与其他饰物的呼应统一。

结的组合方法很多，例如：将结与结钉缝一起；用绳线将结与结穿在一起；在编制过程中利用结的环耳使几个或几种结相互连接；在环耳上编出另外的结；用余线再编出一个或数个新结；等等。

组合结的编制步骤，一般是按先上后下、先外后内、先两边后中间的顺序来完成的。一些在环耳上再编附加结的结，虽主结先编，但编至该环耳处时，须先完成附加结，再完成主结。以环耳相连或以余线相连的结，通常是先编好一结，再在另一结编制的过程中做连接处理。编制组合结一般都是双线头同时进行的，但若碰到单一走向的结和在环耳上附加的结，则需以单线头来编，这时线头走向虽有所改变，但结的编制线路是不变的。

为使花结更加丰富、美观、完整，有时还需搭配一些小的饰物，如珠、环、管子、扣子、坠子等。

usually set up a major knot as visual focus and other parts of the group knot will be connected, extended and stitched. Be sure the neat structure of knots as well as the coherence of color and ornaments in the process of knitting.

There are many ways to combine knots such as sewing and tying them together. Taking advantage of the loop, several knots can be linked up or woven on the loop. Additional knots can be knitted by the remaining thread.

Here are general steps as follows: weaving from the top to the bottom, from the inner side to the outer side, from both ends to the center. Additional knots can be added to loops. Although the main knot is given priority to knit, additional knots have to be done when the loop weaving is about to start. After one of the additional knots is complete, it will be connected with the main knot in the process of knitting another knot. Group knot is generally knitted from both ends of the thread simultaneously; however, one-way knot and knots attached to the loops have to be knitted by a single thread but the knitting route remains unchanged.

Such small ornaments as beads, rings, tubes, buttons and pendants are

这些饰物起到点缀、衬托花结的作用，但不可喧宾夺主。当然，如果需以结来点缀、装饰其他物品的话，则另当别论。

　　结的组合是一种总体设计，既要发挥想象力和创造力，又需要考虑每个结的处理方法，它是一项复杂而有趣味的劳动。

often used to beautify and diversify the knot. All the ornaments serve as a foil to the knot so excessive use will be distracting. If the knot is used to decorate other items, it will be another pair of shoes.

It requires an overall design for the group knot. In addition to imagination and creativity, the layout of the knots should be given consideration. In a word, it is a tough task without loss of fun.

如意结
Ruyi Knot

　　如意是一种象征吉祥的器物，常见以玉、竹、木、骨、金属等材料制成，其头部呈云形或灵芝形，柄部呈"S"形微弯曲，供人观赏或把玩。也有人说如意原为一种挠痒的用具，其弯曲的柄和宽头即为人手的模仿，后宽头被美化为灵芝状的云形，取名为"如意"，即为"可人如意"之意。

　　先编三个相连的三环结，再用三个三环结的余线编一个三环结。

Ruyi is a symbol of good luck made of jade, bamboo, wood, bone, metal and so on. The top resembles glossy ganoderma with cloud pattern. The handle is in "S" shape. Some say that it was a tool for scratching oneself because the design of curved handle and wide top imitated human's hand and arm. To beautify the design, the top is made into glossy ganoderma shape with cloud pattern, named "*Ruyi*" meaning "everything goes as one wishes".

Knit three linked three-loop knots first, and then use the remaining thread to knit a separate three-loop knot.

制作步骤
Knitting Process

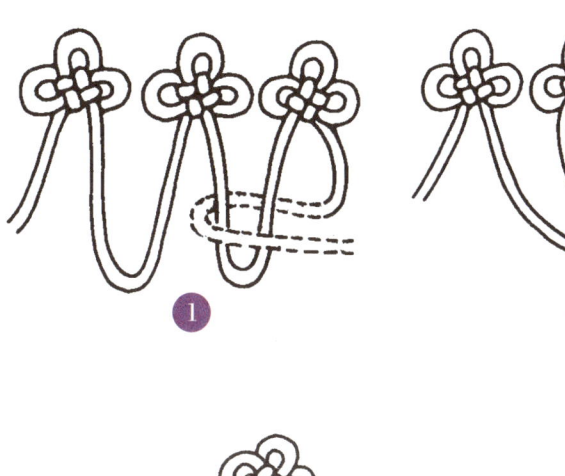

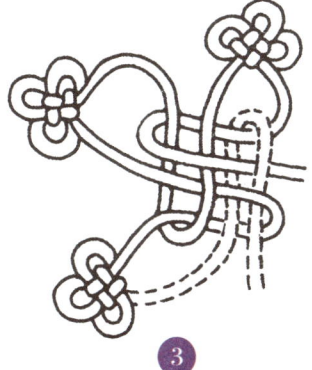

琵琶结
Pipa-shaped Knot

琵琶是一种音色很美的四弦乐器，用木料制成，其下部为瓜子弧形盘（音箱），上部为长柄，柄端弯曲，整体造型流畅饱满。据说琵琶最早来自西域，现为典型的中国传统乐器。

·编一个纽扣结。

·以"扭麻花"的方法编出"琵琶身"。

·扭几个麻花就有几层"V"纹，绳线若较细，可多编几层"V"纹。

·剪去余线再缝几针。

Pipa is a four-stringed musical instrument made of wood with a flowing outline. The top of fretted fingerboard is in long shape with a curved handle and the curved bottom similar to a melon seed. It is said that *Pipa* originally came from the Western Regions (a Han-dynasty term for the area west of Yumenguan, including what is now Xinjiang and parts of Central Asia). It is one of the traditional Chinese musical instruments.

• Knit a button-like knot.

• Fretted fingerboard is like twisting hemp flowers.

• The pattern of "V" comes from "twisting hemp flowers". The more "V" patterns you get, the more "hemp flowers" you have to twister. If the thread is fine enough, more "V" patterns can be woven.

• Cut the remaining thread and then stitch.

制作步骤
Knitting Process

1.
2.
3.
4.
5.
6.
7. 反面 Reverse Side
8. 反面 Reverse Side
9.

109 | Appendix: Knitting Techniques of Several Types of Chinese Knots

附录：几种中国结的编制技法

寿字结
Longevity Character Knot

寿即长命、活得岁数大。人们祈求安康长命、期盼幸福美满，喜欢将福、禄、寿、喜放在一起。在民间寿字有许多种写法，而且常常图案化呈对称格局。寿字也常被作为吉祥符号用于装饰。

- 编一个三环结。
- 用两根余线编两个双环结。
- 用三个结的余线编一个三环结。
- 再编一个三环结。
- 重复编一个三环结和两个双环结。
- 最后以一个三环结收尾。

Happiness, fortune and longevity have been seen as people's common wish so they often appear together. "Longevity" character has a number of variations but the design of "longevity" is mostly symmetrical. As one of the auspicious symbols, "longevity" character knot is often used for decoration.

- Knit a three-loop knot.
- Use the remaining thread to knit two double-loop knots.
- Use the remaining thread to knit a three-loop knot.
- Knit another three-loop knot.
- Knit one more three-loop knot and two double-loop knots.
- End with a three-loop knot.

制作步骤
Knitting Process

① ② ③ ④

111

附录：几种中国结的编制技法
Appendix: Knitting Techniques of Several Types of Chinese Knots

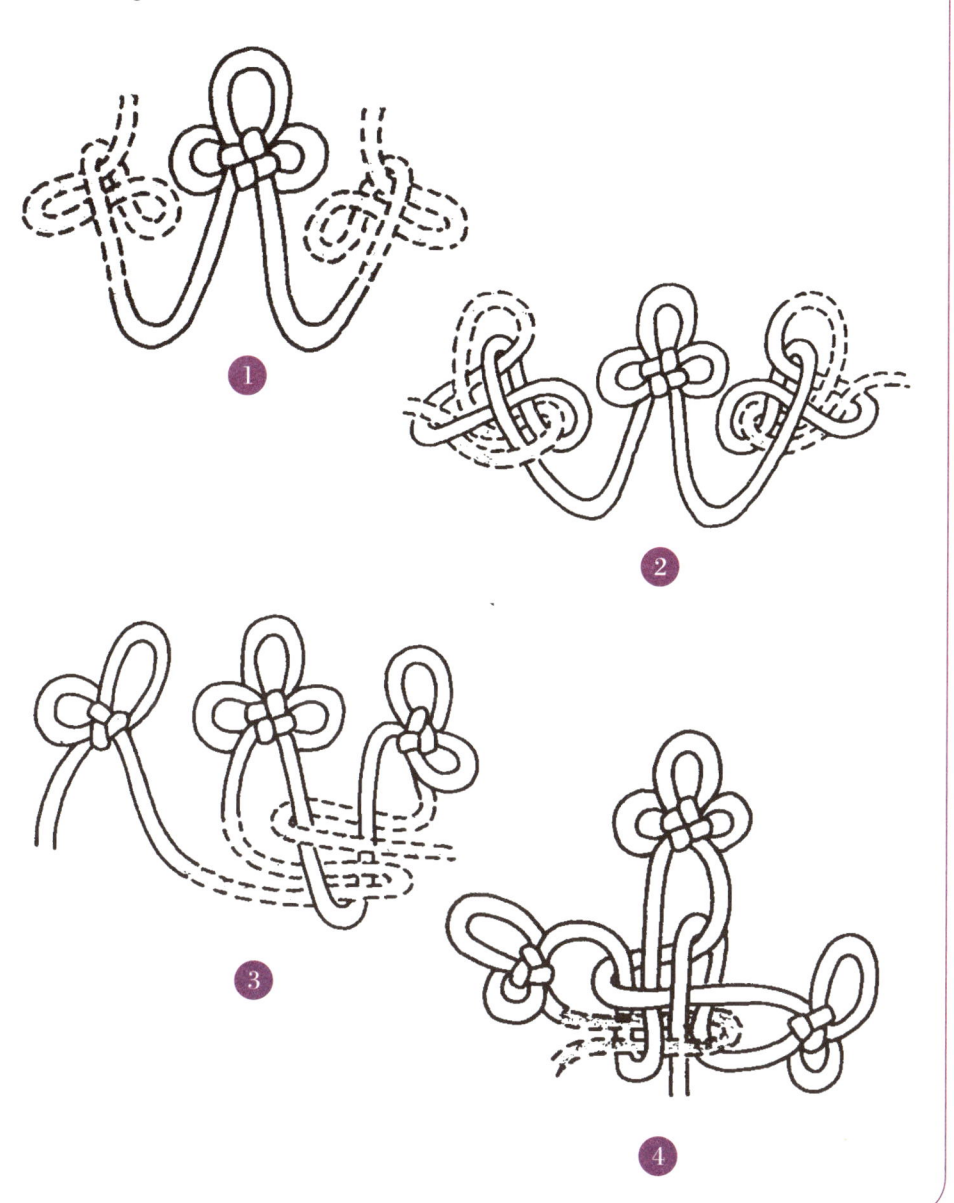

⑤

⑥

⑦

⑧

穗子的编制方法

人们一般喜欢在结的下方做些垂饰或挂上穗子，以增强花结的悬垂感和飘逸感。做垂饰，可在花结的尾部穿些小饰物，或用余线编些小花结，或拆开线头用其他绳线捆扎。穗子是十分常见的垂挂装饰，民间制作的方法很多，在美丽的花结下装上穗子，更增添了一种静与动、点与线、疏与密的对比之美。

The Technique of Weaving Tassels

People love to hang ornaments beneath the knot such as pendants and tassels so as to make the knot look pendulous and graceful. Pendulous accessories can be a small adornment, a small knot knitted by the remaining thread and tassels made of being ripped thrums. Among them, tassels are very common and there are many techniques to make tassels among populace. As a foil, adding tassels highlights the beautiful knot.

同心穗
Concentric Tassels

- 将一束绳线扎牢，再用一线横扎作头。
- 将其反过来用尾线均匀包裹头部，再扎紧即成。

- Fasten a hank of string and lash broadwise as "head".
- Reverse it and wrap up the "head".

113

附录：几种中国结的编制技法
Appendix: Knitting Techniques of Several Types of Chinese Knots

横编穗
Knitting Tassels

此穗是以经纬线编制而成。经线、纬线可同色，也可异色。

　·将奇数的经线依圆筒挂满。

　·将一根纬线横向在经线上作间隔穿梭，间隔规律按不同的设计需要应有所变化。

　·将首尾余线藏进穗头里面。

　• This kind of tassels is woven by warp and weft. They can be either the same color or different colors.

　• Cover the cylinder with odd-numbered warp.

　• Weave weft broadwise into warp at certain intervals. The design of tassels varies with intervals.

　• Hide the loose thread in tassels.

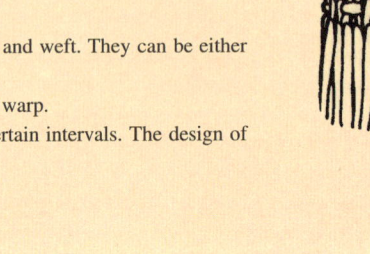

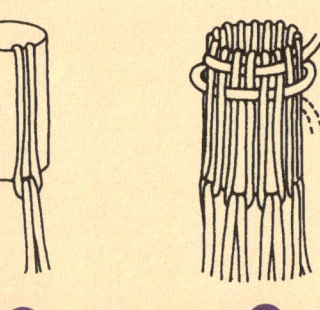

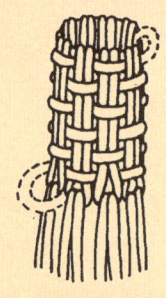

① ② ③

④

• 有结饰的挂饰
Decoration with Knots